# Fortran 90
# Explained

# Fortran 90
# Explained

### MICHAEL METCALF

*Data Handling Division,*
*CERN, Geneva, Switzerland*

### JOHN REID

*Atlas Centre, Rutherford Appleton Laboratory,*
*Oxfordshire, UK*
*(formerly of Harwell Laboratory)*

OXFORD    NEW YORK    TOKYO
OXFORD UNIVERSITY PRESS
1990

Oxford University Press, Walton Street, Oxford OX2 6DP
Oxford New York Toronto
Delhi Bombay Calcutta Madras Karachi
Petaling Jaya Singapore Hong Kong Tokyo
Nairobi Dar es Salaam Cape Town
Melbourne Auckland
and associated companies in
Berlin Ibadan

Oxford is a trade mark of Oxford University Press

Published in the United States
by Oxford University Press, New York

British Library Cataloguing in Publication Data
Metcalf, Michael
Fortran 90 explained
1. Computer systems. Programming
languages. Language Fortran 90.
I. Title.   II. Reid, John K. (John Ker)
1938–
005.133
ISBN 0–19–853772–7

Library of Congress Cataloging in Publication Data
(data applied for)

Printed in Great Britain by
Bookcraft (Bath) Ltd
Midsomer Norton, Avon

# PREFACE

Fortran has always been the principal language used in the fields of scientific, numerical, and engineering programming, and a series of revisions to the standard defining successive versions of the language has progressively enhanced its power and kept it competitive with several generations of rivals.

Since 1978, the technical committee responsible for the development of Fortran standards, X3J3, has been labouring to produce a new, much-needed modern version of the language. Its purpose is to "promote portability, reliability, maintainability, and efficient execution ... on a variety of computing systems". Now this standard is almost ready, and it seems to us appropriate to prepare a definitive informal description of the language it defines. This follows, and is based upon, the first editions of this book, *Fortran 8x Explained*, which described the two drafts of the standard (1987, 1989).

The whole of Fortran 77 is contained in Fortran 90, but certain of its features are labelled 'obsolescent' in the standard, and their use is not recommended. The obsolescent features had replacements already in Fortran 77 and may be removed from the next standard. Certain other features do not have replacements in Fortran 77, but are redundant within Fortran 90. To describe these redundant and obsolescent features in the main body of the text would impair the flow of the exposition, so they have been relegated to a final chapter and an appendix, respectively. It is our hope that as soon as Fortran 90 compilers become available, these features will fall into disuse, and that an understanding of them will be required only when dealing with old programs.

In this book, an initial chapter sets out the background to the work on the new standard, and the nine following chapters describe Fortran 90 less its redundant and obsolescent features in a manner suitable both for grasping the implications of the new features, and for writing programs. Some knowledge of programming concepts, although not necessarily of Fortran 77, is assumed. In order to reduce the number of forward references and also to enable, as quickly as possible, useful programs to be written based on material already absorbed, the order of presentation does not always follow that of the standard.

In order to make the book a complete reference work, it concludes with five appendices. They contain, successively, a list of the intrinsic functions, a summary of Fortran 90 statements, a description of the obsolescent features, a glossary of Fortran terms, and solutions to most of the exercises.

It must be remembered that although the obsolescent features appear in an appendix, they are nevertheless an integral part of the Fortran 90 language. However, the appendix includes advice on how to avoid their use, thereby enhancing the upwards compatibility of programs with respect to possible future standards. The same is true for the features whose use we deprecate, and which are described in Chapter 11.

It is our hope and intention that this book, by providing a complete description of Fortran 90, will play a helpful rôle in the early years of the new standard, as new compilers are introduced and used, and will serve as a long-term reference work throughout the decade and beyond.

# ACKNOWLEDGEMENTS

The development of the Fortran 90 standard has been a long procedure involving several hundred people in many countries. The main burden has fallen on the principal members of X3J3, and especially on its chairman, Jeanne Adams. We extend our thanks to her and all our colleagues on X3J3 and elsewhere for their devotion to this important but thankless task, and for creating such a friendly working atmosphere. We will long remember the week-long meetings held over such an extended period, as well as the personal contacts we made and valued. We have taken great pains to ensure that this book is a true and accurate representation of the final document produced by the committee, and clearly any omissions or other errors or misrepresentations are entirely our responsibility.

It is a pleasure to thank the management of CERN and Harwell, and especially P. Zanella and D.O. Williams of CERN and A.E. Taylor of Harwell, for encouraging us to undertake this work, and for providing the necessary resources for its realisation. JKR is also indebted to the Rutherford Appleton Laboratory for its support in the final months.

We are grateful to Malcolm Cohen (NAG) for checking (and correcting) our examples using a prototype parser, and our final thanks go to Marie-Claire Perler for her helpful cooperation in the preparation of the final camera-ready copy.

# Contents

# 1. WHITHER FORTRAN?

This book is concerned with the Fortran 90 programming language, setting out a reasonably concise description of the whole language. The form chosen for its presentation is that of a textbook intended for use in teaching or learning the language. Its description occupies Chapters 2 to 11, which are written in such a way that simple programs can already be coded after the first three of these chapters (on language elements, expressions and assignments, and control) have been read. Successively more complex programs can be written as the information in each subsequent chapter is absorbed. Chapter 5 describes the important new concept of the module and the many enhancements to procedures, Chapter 6 completes the description of the powerful new array features, Chapter 7 considers the details of specifying data objects and the new derived types, and Chapter 8 details the much expanded set of intrinsic procedures. Chapters 9 and 10 cover the whole of the input/output features in a manner such that the reader can also approach this more difficult area feature by feature, but always with a useful subset already covered. Finally, Chapter 11 describes those features that are redundant in the language, and whose use we choose to deprecate. Here we emphasise that this deprecation represents our own opinion, and not that of X3J3. In a concluding section of each of Chapters 2 to 10, we summarize the differences between Fortran 90 and Fortran 77.

This introductory chapter has the task of setting the scene for those that follow. The first section presents the Fortran language and its considerable evolution since it was first introduced about thirty years ago. The second continues with a justification for preparing a new standard and summarizes the important new features, the third outlines how standards are developed, and the fourth looks at the mechanism that has been proposed to permit the language to evolve in the next few revision cycles. The fifth concludes by considering the requirements on programs and processors for conformance with the standard.

## 1.1 Fortran history

Programming in the early days of computing was tedious in the extreme. Programmers required a detailed knowledge of the instructions, registers, and other aspects of the central processing unit (CPU) of the computer for which they were writing code. The *source code* itself was written in a numerical notation, so called *octal code*. In the course of time mnemonic codes were introduced, a form of coding known as *machine* or *assembly code*. These codes were translated into the instruction words by programs known as *assemblers*. In the 1950s it became increasingly apparent that this form of programming was highly inconvenient, if only because of the length of time required to write and test a program, although it did enable the CPU to be used in a very efficient way.

These difficulties spurred a team led by John Backus of IBM to develop one of the earliest high-level languages, Fortran. Their aim was to produce a language which would be simple to understand but almost as efficient in execution as assembly language. In this they succeeded beyond their wildest dreams. The language was indeed simple to learn, as it was possible to write mathematical formulae almost as they are usually written in mathematical texts. (In fact, the name Fortran is a contraction of Formula Translation.) This enabled working programs to be written faster than before, for only a small loss in efficiency, as a great deal of care was devoted to the construction of the compiler.

But Fortran was revolutionary as well as innovatory. Programmers were relieved of the tedious burden of using assembler language, and were able to concentrate more on the problem in hand. Perhaps more important, however, was the fact that computers became accessible to any scientist or engineer willing to devote a little effort to acquiring a working knowledge of Fortran; no longer was it necessary to be an expert on computers to be able to write application programs.

Fortran spread rapidly as it fulfilled a real need. Inevitably dialects of the language developed, which led to problems in exchanging programs between computers, and so in 1966, after four years' work, the then American Standards Association (later the American National Standards Institute, ANSI) brought out the first ever standard for a programming language, now known as Fortran 66. Essentially, it was a common subset of the dialects, so that each dialect could be regarded as an extension of the standard. Users wishing to write portable code had to be careful to avoid extensions.

Fortran brought with it several other advances, apart from its ease of learning combined with a stress on efficient execution of code. It was, for instance, a language which remained close to, and exploited, the available hardware rather than being an abstract concept. It also brought with it, via the COMMON and EQUIVALENCE statements (see Chapter 11), the possibility for programmers to control storage allocation in a simple way, a feature which was very necessary in those early days of small memories, even if it is now regarded as being potentially dangerous, such that we deprecate its use. Lastly, the source code permitted blank characters in its syntax, freeing the programmer from the duty of writing code in rigidly defined columns and allowing the bodies of the statements to be laid out in any desired way.

The proliferation of dialects remained a problem after the publication of the 1966 standard. A first difficulty was that many compilers did not even adhere to the standard. A second was the widespread implementation in compilers of features which were essential for large-scale programs, but which were ignored by the standard. An example of this was direct-access file-handling. Different compilers implemented such facilities in different ways.

This situation, combined with the existence of some evident flaws in the language, such as the lack of structured programming constructs, resulted in the introduction of large numbers of so-called *pre-processors*. These are programs that are able to read in the source code of some well-defined extended dialect of Fortran and to generate a second text in standard Fortran which is then presented to the Fortran compiler in the usual way. This provided a means for extending Fortran, yet still retaining the ability to transport the transformed source code from one computer to another. At the same time, the large number of such pre-processors meant that there was an even greater diversity of high-level dialects in use. Although programs written using a pre-processor could be exchanged at the Fortran source level, the automatically generated Fortran code was often unacceptably difficult to read.

These difficulties were partially resolved by the publication of a new standard, in 1978, known as Fortran 77. It included several new features that were based on vendor extensions or pre-processors and was, therefore, not a common subset of existing dialects but rather a new dialect in its own right. The transition period between Fortran 66 and Fortran 77 was much longer than it should have been, due to delays in the availability of the new compilers, and the two standards had to coexist for a considerable time. By the mid-1980s, however, the changeover to Fortran 77 was in full swing, and Fortran 66 code was being phased out rapidly. Manufacturers began to stop supporting the old compilers, thus increasing the pressure to change. It was a relatively simple matter to write new code under the new standard and converting old standard-conforming code was usually easy as there is a large measure of compatibility between the two standards. On the other hand, programs that had used extensions were often difficult to convert, as the new compilers often did not include the former extensions. Indeed, some compilers were very strict, implementing Fortran 77 without extensions.

## 1.2  The new standard

After thirty years' existence, Fortran is far from being the only programming language available on most computers. In the course of time new languages have been developed, and where they were demonstrably more suitable for a particular type of application they have been adopted in preference to Fortran for that purpose. Fortran's superiority has always been in the area of numerical, scientific, engineering, and technical applications, and there is still no significant competitor in these fields. The Fortran community has a truly vast investment in Fortran codes, with many programs (some of 100,000 lines or more) in frequent use. This does not mean, however, that the community is necessarily completely content with the language, and in order that it be brought properly up-to-date, the ANSI-accredited technical committee X3J3 has once again prepared a new standard, formerly known as Fortran 8x and now as Fortran 90.

What are the justifications for continuing to revise the definition of the Fortran language? As well as standardizing vendor extensions, there is a need to modernize it in response to the developments in language design which have been exploited in other languages, such as APL, Algol 68, Pascal, and Ada. Here, X3J3 can draw on the obvious benefits of concepts like data hiding. In the same vein is the need to begin to provide an alternative to dangerous storage association, to abolish the rigidity of the outmoded source form, and to improve further on the regularity of the language, as well as to increase the safety of programming in the language and to tighten the conformance requirements. To preserve the vast investment in Fortran 77 codes, the whole of Fortran 77 is contained as a subset.

However, unlike the previous standard, which resulted almost entirely from an effort to standardize *existing practices,* the new standard is much more a *development* of the language, introducing features which are new to Fortran, but are based on experience in other languages. The most significant new features are the ability to handle arrays using a concise but powerful notation, and the ability to define and manipulate user-defined data types. The first of these will lead to a simplification in the coding of many mathematical problems, and will also make Fortran a more efficient language on the new generation of supercomputers as these array features are well matched to their hardware. The second enables programmers to express their problems in terms of data types exactly matched to their requirements.

A summary of the new features is:

i) A means for the language to evolve by labelling some features as 'obsolescent' (see Section 1.4).

ii) Array operations (see Chapter 6).

iii) Pointers.

iv) Improved facilities for numerical computation including a set of numeric inquiry functions.

v) Parametrization of the intrinsic types, to permit processors to support short integers, very large character sets, more than two precisions for real and complex, and packed logicals.

vi) User-defined derived data types composed of arbitrary data structures and operations upon those structures.

vii) Facilities for defining collections called 'modules', useful for global data definitions and for procedure libraries. These support a safe method of encapsulating derived data types.

viii) Requirements on a compiler to detect the use of constructs that do not conform to the syntax of the language or are obsolescent.

ix) A new source form, more appropriate to use at a terminal.

x) New control constructs such as the SELECT CASE construct and a new form of the DO.

xi) The ability to write internal procedures and recursive procedures, and to call procedures with optional and keyword arguments.

xii) Dynamic storage allocation.

xiii) Improvements to the input-output facilities, including handling partial records and a standardized NAMELIST facility.

xiv) Many new intrinsic procedures.

Taken together, the new features contained in Fortran 90 should ensure that the Fortran language will continue to be used successfully for a long time to come. The fact that it contains the whole of Fortran 77 as a subset means that conversion to Fortran 90 will be as simple as conversion to another Fortran 77 processor.

## 1.3 Standardization

The Fortran language has twice been standardized in the framework of ANSI and the International Standards Organization (ISO), in 1966 and 1978. Following the publication of the second standard, the committee responsible for Fortran work, X3J3, re-formed in order to begin work on a third standard, with the intention of having it ready in 1982. This time-scale was hopelessly optimistic; the new language was given the working name Fortran 8x, but as time passed, many could see the 'x' becoming a hexadecimal digit (they were right!).

X3J3 itself is a body composed of about forty-five representatives of computer hardware and software vendors, users, and academia. It is accredited to ANSI, the body that publishes final American standards, but reports directly to its parent committee, X3 (computer systems), which is responsible for actually adopting, or rejecting, the proposed draft standards presented to it. In these decisions, it tries to ensure that the proposals really do represent a consensus of those concerned. The work of X3J3 is tracked by a corresponding international group, ISO/IEC JTC1/SC22/WG5 (which we abbreviate to WG5), consisting of international experts responsible for recommending that a draft standard become an international standard as well as a national one.  X3J3 maintains other close contacts with the international community by having half-a-dozen foreign members, including both of the present authors.

As part of the lengthy standardization procedure, it is necessary to gain at least a two-thirds majority of X3J3 to be able to forward a proposed draft standard to X3, before that committee can consider it.  Once X3 agrees to begin consideration, a period of public comment is initiated, and X3J3 is required to take note of and act on the public comments, and to submit a revised draft to X3.  This cycle can be followed several times.

By 1986, X3J3 was at the fourteenth of the eighteen milestones which have to be passed before a new standard receives full international recognition, and there were strident voices opining that the advances it had made had gone too far, and that there must be some further retrenchment before it was accept-able to them.  A ballot of X3J3 held in April 1986, whose result was 16 to 19 against the draft of that date, led to some slimming down of the language, in an attempt to reach a compromise between radicals who want a large range of new features and conservatives with more modest goals and who emphasize the importance of long-term compatibility.  A second ballot held in January 1987 — 29 to 7 — showed that better agreement had been reached, but that no true consensus had been achieved, given that some of the dissenting votes were those of large vendors claiming to act as representatives of their users. This failure to agree was confirmed in the final ballot to forward the document to X3, held in May 1987: 26 to 9.

Subsequently, X3 approved by a large majority to allow the first draft standard to go to public comment, and over 400 letters were received.  Many shades of opinion were represented in these replies, but a general feeling was that the language was too complex, lacked certain popular extensions, and was burdened by unattractive features.  As a result of this rather sobering experi-ence, a number of members of X3J3 attempted to prepare plans to respond in various ways to this public sentiment, but none succeeded in securing suffi-cient support from the committee.  Finally, in September 1988, WG5, despairing of ever seeing a standard emerge from X3J3, defined exactly which changes it required to the original draft, and set its convenor a timetable to prepare a second draft conforming to its requirements.  Defiantly, it renamed the proposed language Fortran 88!  This action sent shock waves through the standards world, as an international committee was essentially reprimanding a national one and setting it an ultimatum.  WG5's action received support from other bodies, including X3, which sent X3J3 a message reminding it of its international as well as its national obligations.  In response to this pressure, X3J3 voted 24 to 9 in November 1988 to accept the principle of WG5's plan, and at a rather extraordinary meeting passed proposals corresponding, with some differences, to eighteen of WG5's wishes, rejected two, and continued work on the two remaining more difficult ones.  These were completed three months later. In addition, three features not requested by WG5 were added. In May 1989, a second draft proposed standard was forwarded to X3, and was the subject of American and international comment that autumn. This time, about 150 letters were received and many were favourable and expressed impatience to see the project completed. A final draft standard, resulting from changes of a minor nature applied to the second draft as a result of this second period of public comment, is now undergoing final processing as an ISO and a US standard. Unfortunately, the changes are such that a third round of public comment is probable in the US. As a consequence, the ISO standard is likely to be formally adopted much earlier than the US standard.

The dissension within the committee could have been regarded as healthy as long as it led to a constructive debate on its objectives and results in a final resolution of the differences. However, some of the issues were indeed difficult, with complete reconciliation unobtainable:

i) Should Fortran 90 be innovative, or merely standardize existing practice?

ii) Should the language be small and simple, or big and powerful? Is the present proposal too big, making it impossible to fit into the small machines of the 1990s, impossible to be implemented by small software houses, and impossible to be understood by non-professional programmers (who form the bulk of Fortran's users)?

iii) Are users prepared to drop existing features, given 10 to 20 years' notice, or must all existing code work for ever?

iv) Are subsets of the language useful (as a migration aid, for example) or an impediment to portability?

v) Is the proposed architecture for language evolution (see next section) really viable?

vi) Do users want a safe, reliable language, or one which permits them to write the tricky programs more often associated with assembly language programming?

vii) Are the existing proposals difficult and inefficient to implement? Does this matter if users thereby have an easier life?

viii) Will the presence of any of the new features cause any existing ones to be implemented less efficiently?

Now that the various issues are finally resolved, it is our sincere hope that the expectant Fortran community will be provided with working compilers corresponding to a new standard early in the 1990s. However, one hitch in the smooth transition from Fortran 77 to Fortran 90 results from a last-minute decision by X3, in 1989, that within the USA Fortran 90 should be an *additional* standard rather than a *replacement* for Fortran 77. X3J3 was not consulted and expressed its disapproval in a letter ballot. To what extent this will delay and extend the transition period between the two standards cannot yet be foreseen.

## 1.4  Language evolution

The procedures under which X3J3 works require that a period of notice be given before any existing feature is removed from the language. This means, in practice, a minimum of one revision cycle, which for Fortran means a decade or so. The need to remove features is evident: if the only action of the committee is to add new features, the language will become grotesquely large, with many overlapping and redundant items. The solution finally adopted by X3J3 is to publish as an appendix to the draft standard a set of two lists showing which items have been removed or are candidates for eventual removal.

One list contains the *deleted features*, those that have been removed. Since Fortran 90 contains the whole of Fortran 77, this list is empty.

The second list contains the *obsolescent features*, those considered to be already redundant, and which are recommended for deletion in the next revision (although this is not binding on the next committee). The obsolescent features are:

   i)  arithmetic-IF;

  ii)  branching to an END IF statement from outside its block;

 iii)  real and double precision DO variables and control expressions;

  iv)  shared DO termination, and DO termination on a statement other than on a CONTINUE or an END DO statement;

   v)  ASSIGN and the assigned GO TO;

  vi)  alternate RETURN;

 vii)  PAUSE; and

viii)  assigned FORMAT specifiers.

The obsolescent features mechanism should permit a slow reduction in the size of the language in the course of time.

If the Fortran 90 standard is successful, it will surely be subject to further revision, for example to add exception handling and features for parallel processing. One way the language can be readily extended immediately is for modules containing certain features to be written and to be recognised as collateral standards by X3J3. Examples of such modules might be for matrix manipulation and for dynamic character strings. These modules must be written without using obsolescent features. This provides a means of ensuring that Fortran 90 remains a powerful and well-honed tool for numerical and scientific applications for the next decade and beyond.

In the following chapter we begin a description of the Fortran 90 language, but confine the obsolescent features to Appendix C.

## 1.5 Conformance

The standard is almost exclusively concerned with the rules for programs rather than processors. A processor is required to accept a standard-conforming program and to interpret it according to the standard, subject to limits that the processor may impose on the size and complexity of the program. The processor is allowed to accept further syntax and to interpret relationships that are not specified in the standard, provided they do not conflict with the standard. Of course, the programmer must avoid such syntax extensions if portability is desired.

The interpretation of some of the standard syntax is *processor-dependent*, that is, may vary from processor to processor. For example, the set of characters allowed in character strings is processor dependent. Care must be taken whenever a processor-dependent feature is used in case it leads to the program not being portable to a desired processor.

A drawback of the Fortran 77 standard is that it makes no statement about requiring processors to provide a means to detect any departures from the allowed syntax by a program, as long as that departure does not conflict with the syntax rules defined by the standard. The new standard is written in a different style to the old one. The syntax rules are expressed in a form of BNF with associated constraints, and the semantics are described by the text. This semi-formal style is not used in this book, so an example is perhaps helpful:

R609  *substring*  **is** *parent-string(substring-range)*

R610  *parent-string*  **is** *scalar-variable-name*
          **or** *array-element*
          **or** *scalar-structure-component*
          **or** *scalar-constant*

R611  *substring-range*  **is** [*scalar-int-expr*] : [*scalar-int-expr*]

Constraint:  *parent-string* must be of type character.

The first *scalar-int-expr* in *substring-range* is called the **starting point** and the second one is called the **ending point**. The length of a substring is the number of characters in the substring and is MAX(*ending-point* − *starting-point* + 1, 0).

Here, the three production rules and the associated constraint for a character substring are defined, and the meaning of the length of such a substring explained.

The standard is written in such a way that a processor, at compile-time, may check that the program satisfies all the constraints. In particular, the processor must provide a capability to detect and report the use of any

  i) obsolescent feature,

 ii) additional syntax,

iii) kind type parameter (Section 2.5) that it does not support,

iv) non-standard source form or character,

 v) name that is inconsistent with the scoping rules, or

vi) non-standard intrinsic procedure.

Furthermore, it must be able to report the reason for rejecting a program. These capabilities will be of great value in producing correct and portable code. They were not required for Fortran 77 programs.

# 2. LANGUAGE ELEMENTS

## 2.1 Introduction

Written prose in a natural language, such as an English text, is composed firstly of basic elements — the letters of the alphabet. These are combined into larger entities, words, which convey the basic concepts of objects, actions, and qualifications. The words of the language can be further combined into larger units, phrases and sentences, according to certain rules. One set of rules defines the grammar. This tells us whether a certain combination of words is correct in that it conforms to the *syntax* of the language, that is those acknowledged forms which are regarded as correct renderings of the meanings we wish to express. Sentences can in turn be joined together into paragraphs, which conventionally contain the composite meaning of their constituent sentences, each paragraph expressing a larger unit of information. In a novel, sequences of paragraphs become chapters and the chapters together form a book, which usually is a self-contained work, largely independent of all other books.

## 2.2 Fortran character set

Analogies to these concepts are found in a programming language. In Fortran 90, the basic elements, or character set, are the 26 upper- and lower-case letters of the English alphabet, the 10 Arabic numerals, 0 to 9, the underscore, _, and the so-called special characters listed in Table 1. Within the Fortran 90 syntax, the upper- and lower-case letters are equivalent; they are distinguished only when they form part of character sequences. In this book, syntactically significant characters will always be written in upper case. The letters, numerals, and underscore are known as *alphanumeric* characters.

Except for the currency symbol, whose graphic may vary (for example, to be £ in the United Kingdom), the graphics are fixed, though their styles are not fixed.

Table 1. The special characters of the Fortran 90 language.

| Character | Name | Character | Name |
|---|---|---|---|
| = | Equals sign | : | Colon |
| + | Plus sign |   | Blank |
| − | Minus sign | ! | Exclamation mark |
| * | Asterisk | " | Quotation mark |
| / | Slash | % | Percent |
| ( | Left parenthesis | & | Ampersand |
| ) | Right parenthesis | ; | Semicolon |
| , | Comma | < | Less than |
| . | Decimal point | > | Greater than |
| $ | Currency symbol | ? | Question mark |
| ' | Apostrophe | | |

In the course of this and the following chapters, we shall see how further analogies with natural language may be drawn. The unit of Fortran 90 information is the *lexical token*, which corresponds to a word or punctuation mark. Adjacent tokens are usually separated by spaces or the end of a line, but sensible exceptions are allowed just as for a punctuation mark in prose. Sequences of tokens form *statements*, corresponding to sentences. Statements, like sentences, may be joined to form larger units like paragraphs. In Fortran 90 these are known as *program units*, and out of these may be built a *program*. A program forms a complete set of instructions to a computer to carry out a defined sequence of operations. The simplest program may consist of only a few statements, but programs of more than 100,000 statements are now quite common.

## 2.3 Tokens

Within the context of Fortran 90, alphanumeric characters (the letters, the underscore, and the numerals) may be combined into sequences that have one or more meanings. For instance, one of the meanings of the sequence 999 is a constant in the mathematical sense. Similarly, the sequence DATE might represent, as one possible interpretation, a variable quantity to which we assign the calendar date.

The special characters are used to separate such sequences and also have various meanings. We shall see how the asterisk is used to specify the operation of multiplication, as in X*Y, and has also a number of other interpretations.

Basic significant sequences of alphanumeric characters or of special characters are referred to as *tokens;* they are labels, keywords, names, constants (other than complex literal constants), operators, and *separators*, which are

```
/   (   )   (/   /)   ,   =   =>   :   ::   ;   %
```

For example, the expression X*Y contains the three tokens X, *, and Y.
Apart from within a character string or within a token, blanks may be used
freely to improve the layout. Thus, whereas the variable DATE may not be
written as D A T E, X * Y is syntactically equivalent to X*Y. In this
context, multiple blanks are syntactically equivalent to a single blank.

A name, constant, or label must be separated from an adjacent keyword,
name, constant or label by one or more blanks or by the end of a line. For
instance, in

```
      REAL X
      READ 10
  30  DO K=1,3
```

the blanks are required after REAL, READ, 30, and DO. However, some
pairs of keywords, such as ELSE IF, are not required to be separated. Similarly, some keywords may be split; for example INOUT may be written
IN OUT. We do not use these alternatives in the main text, but the exact
rules are given in the statement summaries in Appendix B.

## 2.4  Source form

Fortran 90 brings with it a new source form, well adapted to use at a terminal.
The statements of which a source program is composed are written on *lines*.
Each line may contain up to 132 characters,[1] and usually contains a single
statement. Since leading spaces are not significant, it is possible to start all
such statements in the first character position, or in any other position consistent with the user's chosen layout. A statement may thus be written as

```
  X = (-Y + ROOT_OF_DISCRIMINANT)/(2.0*A)
```

In order to be able to mingle suitable comments with the code to which
they refer, Fortran 90 allows any line to carry a trailing comment field, following an exclamation mark (!). An example is

```
  X = Y/A - B    ! Solve the linear equation
```

---

[1] Lines containing characters of nondefault kind (Sections 2.6.4) are subject to a
processor-dependent limit.

Any comment always extends to the end of the source line and may include processor-dependent characters (it is not restricted to the Fortran character set, Section 2.2). Any line whose first non-blank character is an exclamation mark, or contains only blanks, or which is empty, is purely commentary, and is ignored by the compiler. Such comment lines may appear anywhere in a program unit, including ahead of the first statement (but not after the final program unit). A *character context* (those contexts defined in Sections 2.6.4, 9.13.2, 9.13.3, and C.7) is allowed to contain !, so the ! does not initiate a comment in this case; in all other cases it does.

Since it is possible that a long statement might not be accommodated in the 132 positions allowed in a single line, up to 39 additional continuation lines are allowed. The so-called *continuation mark* is the ampersand (&) character, and this is appended to each line that is followed by a continuation line. Thus, the first statement of this section (considerably spaced out) could be written as

```
X =                                           &
   (-Y + ROOT_OF_DISCRIMINANT)                &
   /(2.0*A)
```

On a noncomment line, if & is the last non-blank character or the last non-blank character ahead of the comment symbol !, the statement continues from the character immediately preceding the &. Normally, continuation is to the first character of the next noncomment line, but if the first non-blank character of the next noncomment line is &, continuation is to the character following the &. For instance, the above statement may be written

```
X =                                           &
   &(-Y + ROOT_OF_DISCRIMINANT)/(2.0*A)
```

In particular, if a token cannot be contained at the end of a line, the first non-blank character on the next noncomment line must be an & followed immediately by the remainder of the token.

Comments are allowed to contain any characters, including &, so they cannot be continued since a trailing & would be taken as part of the comment. However, comment lines may be freely interspersed among continuation lines and do not count towards the limit of 39 lines.

In a character context, continuation must be from a line without a trailing comment and to a line with a leading ampersand. This is because both ! and & are permitted both in character contexts and in comments.

No line is permitted to have & as its only non-blank character, or as its only non-blank character ahead of !. Such a line is really a comment and becomes a comment if & is removed. In this book, the ampersands will normally be aligned to improve readability.

When writing short statements one after the other, it can be convenient to write several of them on one line.  The semi-colon (;) character is used as a *statement separator* in these circumstances, for example:

```
A = 0; B = 0; C = 0
```

Since commentary always extends to the end of the line, it is not possible to insert commentary between statements on a single line.  In principle, it is possible to write even long statements one after the other in a solid block of lines, each 132 characters long and with the appropriate semi-colons separating the individual statements.  In practice, such code is unreadable, and the use of multiple-statement lines should be reserved for trivial cases such as the one shown in this example.

Any Fortran 90 statement (that is not part of a compound statement) may be labelled, in order to be able to identify it.  For some statements a label is mandatory.  A statement *label* precedes the statement, and is regarded as a token.  The label consists of from one to five digits, one of which must be non-zero.  An example of a labelled statement is

```
100 CONTINUE
```

Leading zeros are not significant in distinguishing between labels.  For example, 10 and 010 are equivalent.

## 2.5  Concept of type

In Fortran 90, it is possible to define and manipulate various types of data. For instance, we may have available the value 10 in a program, and assign that value to an integer scalar variable denoted by I.  Both 10 and I are of type integer; 10 is a fixed or *constant* value, whereas I is a *variable* which may be assigned other values.  Integer expressions, such as I+10, are available too.

A *data type* consists of a set of data values, a means of denoting those values, and a set of operations that are allowed on them.  For the integer data type, the values are ..., $-3$, $-2$, $-1$, 0, 1, 2, 3,...  between some limits depending on the kind of integer and computer system being used.  Such tokens as these are *literal constants*, and each data type has its own form for expressing them.  Named scalar variables, such as I, may be established. During the execution of a program, the value of I may change to any valid value, or may become *undefined*, that is have no predictable value.  The operations which may be performed on integers are those of usual arithmetic; we can write 1+10 or I $-$ 3 and obtain the expected results.  Named constants may be established too; these have values that do not change during a given execution of the program.

Properties like those just mentioned are associated with all the data types of Fortran 90, and will be described in detail in this and the following chapters. The language itself contains five data types whose existence may always be assumed. These are known as the *intrinsic data types*, whose literal constants form the subject of the next section. Of each intrinsic type there is a default kind and a processor-dependent number of other kinds. Each kind is associated with a nonnegative integer value known as the *kind type parameter*. This is used as a means of identifying and distinguishing the various kinds available.

In addition, it is possible to define other data types based on collections of data of the intrinsic types, and these are know as *derived data types*. The ability to define data types of interest to the programmer — matrices, geometrical shapes, lists, interval numbers — is a powerful feature of the language, one which permits a high level of *data abstraction*, that is the ability to define and manipulate data objects without being concerned about their actual representation in a computer.

## 2.6  Literal constants of intrinsic type

The intrinsic data types are divided into two classes. The first class contains the three *numeric* types which are used mainly for numerical calculations — integer, real, and complex. The second class contains the two *non-numeric* types which are used for such applications as text-processing and control - character and logical. The numerical types are used in conjunction with the usual operators of arithmetic, such as + and −, which will be described in Chapter 3. Each includes a zero and the value of a signed zero is the same as that of an unsigned zero. The non-numeric types are used with sets of operators specific to each type; for instance, character data may be concatenated. These too will be described in Chapter 3.

## 2.6.1  Integer literal constants

The first type of literal constant is the *integer literal constant*. The default kind is simply a signed or unsigned integer value, for example

```
1
0
-999
32767
+10
```

The *range* of the default integers is not specified in the language, but on a computer with a word size of $n$ bits, is often from $-2^{n-1}$ to $+2^{n-1}-1$. Thus on a 16-bit computer the range is often from $-32768$ to $+32767$.

To be sure that the range will be adequate on any computer requires the specification of the kind of integer by giving a value for the kind parameter. This is best done through a named integer constant. For example, if the range $-999999$ to $999999$ is desired, K6 may be established as a constant with an appropriate value by the statement, fully explained later,

```
INTEGER, PARAMETER :: K6=SELECTED_INT_KIND(6)
```

and used in constants thus:

```
-1234567_K6
+1_K6
 2_K6
```

Here, SELECTED_INT_KIND(6) is an intrinsic inquiry function call, and it returns a kind parameter value that yields the range $-999999$ to $999999$ with the least margin (see Section 8.7.4).

On a given processor, it might be known that the kind value needed is 3. In this case, the first of our constants can be written

```
-1234567_3
```

but this form is less portable. If we move the code to another processor, this particular value may be unsupported, or might correspond to a different range.

We expect many implementations to use kind values that indicate the number of bytes of storage occupied by a value, but the standard allows greater flexibility. For example, a processor might have hardware only for 4-byte integers, and yet support kind values 1, 2, and 4 with this hardware (to ease portability from processors that have hardware for 1-, 2-, and 4-byte integers). However, the standard makes no statement about kind values or their order, except that the kind value is never negative.

The value of the kind type parameter for a given data type on a given processor can be obtained from the KIND intrinsic function (Section 8.2):

```
KIND(1)      for the default value
KIND(2_K6)   for the example
```

and the actual range of a given entity may be obtained from another function (Section 8.7.2), as in

```
RANGE(2_K6)
```

which in this case would return a value of at least 6.

In addition to the usual integers of the decimal number system, for some applications it is very convenient to be able to represent positive whole numbers in binary, octal, or hexadecimal form. Unsigned constants of these forms exist in Fortran 90, and are represented as illustrated in these examples:

```
binary (base 2):        B'01100110'
octal (base 8):         0'076543'
hexadecimal (base 16):  Z'10FA'
```

In the hexadecimal form, the letters A to F represent the values beyond 9; they may be used also in lower case. The delimiters may be quotation marks or apostrophes. The use of these forms of constants is limited to their appearance as implicit integers in the DATA statement (Section 7.5). A binary, octal, or hexadecimal constant may also appear in an internal or external file (Section 9.3.2) as a digit string, without the leading letter and the delimiters.

## 2.6.2  Real literal constants

The second type of literal constant is the *real literal constant*. The default kind is a floating-point form built of some or all of: a signed or unsigned integer part, a decimal point, a fractional part, and a signed or unsigned exponent part. One or both of the integer part and fractional part must be present. The exponent part is either absent or consists of the letter E followed by a signed or unsigned integer. One or both of the decimal point and the exponent part must be present. An example is

```
-10.6E-11
```

meaning $-10.6 \times 10^{-11}$, and other legal forms are

```
1.
-0.1
1E-1
3.141592653
```

The default real literal constants are representations of a subset of the real numbers of mathematics, and the standard specifies neither the allowed range of the exponent nor the number of significant digits represented by the processor. Common values are about $10^{-38}$ to $10^{+38}$ for the range, with a precision of about seven decimal digits.

To be sure to obtain a desired range and significance, requires the specification of a kind parameter value. For example,

```
INTEGER, PARAMETER :: LONG = SELECTED_REAL_KIND(9, 99)
```

ensures that the constants

```
1.7_LONG
12.3456789E30_LONG
```

have a precision of at least nine significant decimals, and an exponent range of at least $10^{-99}$ to $10^{+99}$. The number of digits specified in the significand has no effect on the kind. In particular, it is permitted to write more digits than the processor can in fact use.

As for integers, we expect many implementations to use kind values that indicate the number of bytes of storage occupied by a value, but the standard allows greater flexibility. It specifies only that the kind value is never negative. If the desired kind value is known it may be used directly, as in the case

```
1.7_4
```

but the resulting code is then less portable.

The processor must provide at least one representation with more precision than the default, and this second representation may also be specified as DOUBLE PRECISION. We defer the description of this alternative but outmoded syntax to Section 11.4.2.

The KIND function is valid also for real values:

```
KIND(1.0)         for the default value
KIND(1.0_LONG)    for the example
```

In addition, there are two inquiry functions available which return the actual precision and range, respectively, of a given real entity (see Section 8.7.2). Thus, the value of

```
PRECISION(1.7_LONG)
```

would be at least 9, and the value of

```
RANGE(1.7_LONG)
```

would be at least 99.

### 2.6.3 Complex literal constants

Fortran 90, as a language intended for scientific and engineering calculations, has the advantage of having as third literal constant type the *complex literal constant*. This is designated by a pair of literal constants, which are either integer or real, separated by a comma and enclosed in parentheses. Examples are

```
(1., 3.2)
(1, .99E-2)
(1.0, 3.7_8)
```

where the first constant of each pair is the real part of the complex number, and the second constant is the imaginary part. If one of the parts is integer, the kind of the complex constant is that of the other part. If both parts are integer, the kind of the constant is that of the default real type. If both parts are real and of the same kind, this is the kind of the constant. If both parts are real and of different kinds, the kind of the constant is that of the part with the greater decimal precision, or is processor dependent if the decimal precisions are the identical.

A default complex constant is one whose kind value is that of default real.

The KIND, PRECISION, and RANGE functions are equally valid for complex entities.

Note that if an implementation uses the number of bytes needed to store a real as its kind value, the number of bytes needed to store a complex value of the corresponding kind is twice the kind value. For example, if the default real type has kind 4 and needs four bytes of storage, the default complex type has kind 4 but needs eight bytes of storage.

### 2.6.4 Character literal constants

The fourth type of literal constant is the *character literal constant*. The default kind consists of a string of characters enclosed in a pair of either apostrophes or quotation marks, for example

```
'ANYTHING GOES'

"NUTS & BOLTS"
```

The characters are not restricted to the Fortran set (Section 2.2). Any graphic character supported by the processor is permitted, but not control characters such as "newline". They may be in upper- and lower-case on a processor that supports both. The apostrophes and quotation marks serve as *delimiters,* and are not part of the value of the constant. The value of the constant

```
'STRING'
```

is STRING.  We note that in character constants the blank character is significant.  For example

```
'A STRING'
```

is not the same as

```
'ASTRING'
```

A problem arises with the representation of an apostrophe or a quotation
mark in a character constant.   Delimiter characters of one sort may be
embedded in a string delimited by the other, as in the examples

```
'HE SAID "HELLO"'
"THIS CONTAINS AN ' "
```

Alternatively, a doubled delimiter without any embedded intervening blanks is
regarded as a single character of the constant.  For example

```
'ISN''T IT A NICE DAY'
```

has the value ISN'T IT A NICE DAY.
    The number of characters in a string is called its *length,* and may be zero.
For instance, '' and "" are character constants of length zero.
    We mention here the particular rule for the source form concerning char-
acter constants that are written on more than one line (needed because constants may include the characters ! and &): not only must each line that is
continued be without a trailing comment, but each continuation line must
*begin* with a continuation mark.  Any blanks following a trailing ampersand or
preceding a leading ampersand are not part of the constant, nor are the ampersands themselves part of the constant.  Everything else, including blanks, is
part of the constant.  An example is

```
LONG_STRING =                                               &
       'Were I with her, the night would post too soon;     &
       & But now are minutes added to the hours;            &
       & To spite me now, each minute seems a moon;         &
       & Yet not for me, shine sun to succour flowers!      &
       &   Pack night, peep day; good day, of night now borrow:   &
       &   Short, night, to-night, and length thyself tomorrow.'
```

On any computer, the characters have a property known as their *collating sequence*. One may ask the question whether one character occurs before or after another in the sequence. This question is posed in a natural form such as 'Does C precede M?', and we shall see later how this may be expressed in Fortran 90 terms. Fortran 90 requires the computer's collating sequence to satisfy the following conditions:

- A is less than B is less than C.... is less than Y is less than Z;

- 0 is less than 1 is less than 2.... is less than 8 is less than 9;

- blank is less than A and Z is less than 0, or blank is less than 0 and 9 is less than A;

and, if the lower-case letters are available,

- a is less than b is less than c.... is less than y is less than z;

- blank is less than a and z is less than 0, or blank is less than 0 and 9 is less than a.

Thus we see that there is no rule about whether the numerals precede or succeed the letters, nor about the position of any of the special characters or the underscore, apart from the rule that blank precedes both partial sequences. Any given computer system has a complete collating sequence, but no program should ever depend on any ordering beyond that stated above. Additionally, Fortran 90 provides access to the collating sequence of the ASCII standard through intrinsic functions (Section 8.5), but this access is not so convenient and is less efficient on some computers.

A processor is required to provide access to the default kind of character constant just described. In addition, it may support other kinds of character constants, in particular those of non-European languages, which may have more characters than can be provided in a single byte. For example, a processor might support Kanji with the kind parameter value 2; in this case a Kanji character constant may be written

        2_'国内'

or

        KANJI_'標準'

where KANJI is an integer constant with the value 2. We note that in this case, the kind type parameter exceptionally *precedes* the constant. This is necessary in order to enable compilers to parse statements simply.

There is no requirement on a processor to provide more than one kind of character, and the standard does not require any particular relationship between the kind parameter values and the character sets and the number of bytes needed to represent them.  In fact, all that is required is that each kind of character set includes a blank character.  As for the other data types, the KIND function gives the actual value of the kind type parameter, as in

```
KIND('ASCII')
```

## 2.6.5 Logical literal constants

The fifth type of literal constant is the *logical literal constant*.  The default kind has one of two values, .TRUE. and .FALSE. .  These logical constants are normally used only to initialize logical variables to their required values, as we shall see in Section 3.6.

The default kind has a kind parameter value which is processor dependent. The actual value is available as KIND(.TRUE.).  As for the other intrinsic types, the kind parameter may be specified by an integer constant following an underscore, as in

```
.FALSE._1
.TRUE._LONG
```

Nondefault logical kinds are useful for storing logical arrays compactly; we defer further discussion until Section 6.14.

## 2.7  Names

A Fortran 90 program references many different entities by name.  Such names must consist of between 1 and 31 alphanumeric characters (letters, underscores, and numerals) of which the first must be a letter.  There are no other restrictions on the names; in particular there are no reserved words in Fortran 90.  We thus see that valid names are, for example,

```
A
A_THING
X1
MASS
Q123
REAL
TIME_OF_FLIGHT
```

and invalid names are

```
1A          First character is not alphabetic
A THING     Contains a blank
$SIGN       Contains a non-alphanumeric character
```

Within the constraints of the syntax, it is important for program clarity to choose names which have a clear significance — these are known as *mnemonic names*. Examples are DAY, MONTH, and YEAR, for variables to store the calendar date.

The use of names to refer to constants, already met in Section 2.6.1, will be fully described in Section 7.4.

## 2.8 Scalar variables of intrinsic type

We have seen in the section on literal constants that there exist five different intrinsic data types. Each of these types may have variables too. The simplest way by which a variable may be declared to be of a particular type is by specifying its name in a *type declaration statement* such as

```
INTEGER I
REAL A
COMPLEX CURRENT
LOGICAL PRAVDA
CHARACTER LETTER
```

Here all the variables have default kind, and LETTER has default length, which is 1. Explicit requirements may also be specified through *type parameters*, as in the examples

```
INTEGER(KIND=4) I
REAL(KIND=LONG) A
CHARACTER(LEN=20, KIND=1) ENGLISH_WORD
CHARACTER(LEN=20, KIND=KANJI) KANJI_WORD
```

Character is the only type to have two parameters, and here the two character variables each have length 20. Where appropriate, just one of the parameters may be specified, leaving the other to take its default value, as in the cases

```
CHARACTER(KIND=KANJI) KANJI_LETTER
CHARACTER(LEN=20) ENGLISH_WORD
```

The shorter forms

```
INTEGER(4) I
REAL(LONG) A
CHARACTER(20, 1) ENGLISH_WORD
CHARACTER(20, KANJI) KANJI_WORD
CHARACTER(20) ENGLISH_WORD
```

are available, but note that

```
CHARACTER(KANJI) KANJI_LETTER          ! Beware
```

is not an abbreviation for

```
CHARACTER(KIND=KANJI) KANJI_LETTER
```

because a single unnamed parameter is taken as the length parameter.

## 2.9  Derived data types

When programming, it is often useful to be able to manipulate objects that are more sophisticated than those of the intrinsic types. Imagine, for instance, that we wished to specify objects representing persons. Each person in our application is distinguished by a name, an age, and an identification number. Fortran 90 allows us to define a corresponding data type in the following fashion:

```
TYPE PERSON
    CHARACTER(LEN=10) NAME
    REAL AGE
    INTEGER ID
END TYPE PERSON
```

This is the *definition* of the type. A scalar object of such a type is called a *structure*. In order to create a structure of that type, we write an appropriate type declaration statement, such as

```
TYPE(PERSON) YOU
```

The scalar variable YOU is then a composite object of type PERSON containing three separate components, one corresponding to the name, another to the age, and a third to the identification number. As will be described in Sections 3.8 an 3.9, a variable such as YOU may appear in expressions and assignments involving other variables or constants of the same or different types. In addition, each of the components of the variable may be referenced individually using the *component selector* character percent (%). The identification number of YOU would, for instance, be accessed as

```
YOU%ID
```

and this quantity is an integer variable which could appear in an expression such as

```
YOU%ID + 9
```

Similarly, if there were a second object of the same type:

```
TYPE(PERSON) ME
```

the differences in ages could be established by writing

```
YOU%AGE - ME%AGE
```

It will be shown in Section 3.8 how a meaning can be given to an expression such as

```
YOU - ME
```

Just as the intrinsic data types have associated literal constants, so too may literal constants of derived type be specified. Their form is the name of the type followed by the constant values of the components, in order and enclosed in parentheses. Thus, the constant

```
PERSON( 'Smith', 23.5, 2541)
```

may be written assuming the derived type defined at the beginning of this section, and could be *assigned* to a variable of the same type:

```
YOU = PERSON( 'Smith', 23.5, 2541)
```

Any such *structure constructor* can appear only after the definition of the type.
   A derived type may have a component that is of a previously defined derived type. This is illustrated in Figure 1. A variable of type TRIANGLE may be declared thus

```
TYPE(TRIANGLE) T
```

and T has components T%A, T%B, and T%C all of type POINT, and T%A has components T%A%X and T%A%Y of type real.

```
TYPE POINT
   REAL X, Y
END TYPE POINT
TYPE TRIANGLE
   TYPE(POINT) A, B, C
END TYPE TRIANGLE
```

Figure 1.

## 2.10  Arrays of intrinsic type

Another compound object supported by Fortran 90 is the *array*. An array consists of a rectangular set of elements, all of the same type and type parameters. There are a number of ways in which arrays may be declared; for the moment we shall consider only the declaration of arrays of fixed sizes. To declare an array named A of 10 real elements, we might use the statement

```
REAL, DIMENSION(10) :: A
```

The successive elements of the array are A(1), A(2), A(3),..., A(10). The number of elements of an array is called its *size*. Each array element is a scalar.

Many problems require a more elaborate declaration than one in which the first element is designated 1, and it is possible in Fortran 90 to declare a lower as well as an upper *bound*:

```
REAL, DIMENSION(-10:5) :: VECTOR
```

This is a vector of 16 elements, VECTOR(–10), VECTOR(–9),..., VECTOR(5). We thus see that whereas we always need to specify the upper bound, the lower bound is optional, and by default has the value 1. The number of elements along a dimension of an array is known as the *extent* in that dimension. Thus, VECTOR has an extent of 16.

An array may extend in more than one dimension, and Fortran 90 allows up to seven dimensions to be specified. For instance

```
REAL, DIMENSION(5,4) :: B
```

declares an array with two dimensions, and

```
REAL, DIMENSION(-10:5, -20:-1, 0:1, -1:0, 2, 2, 2) :: GRID
```

declares seven dimensions, the first four with explicit lower bounds. It may be seen that the size of this second array is 16×20×2×2×2×2×2 = 10240, and that arrays of many dimensions can thus place large demands on the memory of a computer. The number of dimensions of an array is known as its *rank*. Thus, GRID has a rank of seven. Scalars are regarded as having rank zero. The sequence of extents is known as the *shape*. For example, GRID has the shape (16, 20, 2, 2, 2, 2, 2).

A derived type may contain an array component. For example, the following type

```
TYPE TRIPLET
   REAL U
   REAL, DIMENSION(3)   :: DU
   REAL, DIMENSION(3,3) :: D2U
END TYPE TRIPLET
```

might be used to hold the value of a variable in three dimensions and the values of its first and second derivatives. If T is of type TRIPLET, T%DU and T%D2U are arrays of type real.

Some statements treat the elements of an array one-by-one in a  special order which we call the *array element order*. It is obtained by counting most rapidly in the early dimensions. Thus, the first element of GRID is GRID(−10, −20, 0, −1, 1, 1, 1)     and     this     is     followed     by GRID(−9, −20, 0, −1, 1, 1, 1).     The     last     element     of     GRID     is GRID(5, −1, 1, 0, 2, 2, 2).  This is illustrated for an array of two dimensions in Figure 2.  Most implementations actually store arrays in contiguous storage in array element order, but we emphasise that the standard does not require this.

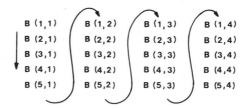

Figure 2. The ordering of elements in the array B(5,4).

We reference an individual element of an array by specifying, as in the examples above, its *subscript* values. In the examples we used integer constants, but in general each subscript may be formed of a *scalar integer expression,* that is, any arithmetic expression whose value is scalar and of type integer. Each subscript must be within the corresponding ranges defined in the array declaration and the number of subscripts must equal the rank. Examples are

```
A(1)
A(I*J)           ! I and J are of type integer
A(NINT(X+3.))    ! X is of type real
T%D2U(I+1,J+2)   ! T is of derived type TRIPLET
```

where NINT is an intrinsic function to convert a real value to the nearest integer (see Section 8.3.1). In addition subarrays, called *sections*, may be referenced by specifying a range for one or more subscripts. The following are examples of array sections:

```
A(I:J)           ! Rank-one array of size J–I+1
B(K, 1:N)        ! Rank-one array of size N
C(1:I, 1:J, K)   ! Rank-two array with extents I and J
```

We describe array sections in more detail in Section 6.10. An array section is itself an array, but its individual elements must not be accessed through the section designator. Thus, B(K, 1:N)(L) cannot be written; it must be expressed as B(K, L).

A further form of subscript is shown in

```
A(IPOINT)        ! IPOINT is an integer array
```

where IPOINT is an array of indices, pointing to array elements. It may thus be seen that A(IPOINT), which identifies as many elements of A as IPOINT has elements, is an example of another *array-valued object*, and IPOINT is referred to as a *vector subscript*. This will be met in greater detail in Section 6.10.

It is often convenient to be able to define an array constant. In Fortran 90, a rank-one array may be constructed as a list of elements enclosed between the tokens (/ and /). A simple example is

```
(/ 1, 2, 3, 5, 10 /)
```

which is an array of rank one and size five. To obtain a series of values, the individual values may be defined by an expression that depends on an integer variable having values in a range, with an optional stride. Thus, the constructor

```
(/1, 2, 3, 4, 5/)
```

can be written as

```
(/ (I, I = 1,5) /)
```

and

```
(/2, 4, 6, 8/)
```

as

```
(/ (I, I = 2,8,2) /)
```

and

```
(/ 1.1, 1.2, 1.3, 1.4, 1.5 /)
```

as

```
(/ (I*0.1, I=11,15) /)
```

An array constant of rank greater than one may be constructed by using the function RESHAPE (see Section 8.13.3) to reshape a rank-one array constant.

A full description of array constructors is reserved for Section 6.13.

## 2.11  Character substrings

It is possible to build arrays of characters, just as it is possible to build arrays of any other type:

```
CHARACTER, DIMENSION(80) :: LINE
```

declares an array, called  LINE, of 80 elements, each one character in length. Each character may be addressed by the usual reference, LINE(I) for example. In this case, however, a more appropriate declaration might be

```
CHARACTER(LEN=80) LINE
```

which declares a scalar data object of 80 characters.  These may be referenced individually or in groups using a *substring* notation

```
LINE(I:J)    ! I and J are of type integer
```

which references all the characters from I to J in LINE. The colon is used to separate the two substring subscripts, which may be any scalar integer expressions. The colon is obligatory in substring references, so that referencing a single character requires LINE(I:I). There are default values for the substring subscripts. If the lower one is omitted, the value 1 is assumed; if the upper one is omitted, a value corresponding to the character length is assumed. Thus,

```
LINE(:I)    is equivalent to LINE(1:I)
LINE(I:)    is equivalent to LINE(I:80)
LINE(:)     is equivalent to LINE or LINE(1:80)
```

If I is greater than J in LINE(I:J), the value is a zero-sized string.

We may now combine the length declaration with the array declaration to build arrays of character objects of specified length, as in

```
CHARACTER(LEN=80), DIMENSION(60) :: PAGE
```

which might be used to define storage for the characters of a whole page, with 60 elements of an array, each of length 80. To reference the line J on a page we may write PAGE(J), and to reference the Ith character on that line we could combine the array subscript and character substring notations into

```
PAGE(J)(I:I)
```

A substring of a character constant or of a structure component may also be formed:

```
'ABCDEFGHIJKLMNOPQRSTUVWXYZ'(J:J)
PERSON%NAME(1:2)
```

At this point we must note a limitation associated with character variables, namely that character variables must have a declared maximum length, making it impossible to manipulate character variables of variable length, unless they are defined appropriately as a derived data type.[2] Nevertheless, this data type is adequate for most character manipulation applications.

---

[2] There is an activity within ISO to define a suitable standard module.

## 2.12  Objects and subobjects

We have seen that derived types may have components that are arrays, as in

```
TYPE TRIPLET
   REAL, DIMENSION(3) :: VERTEX
END TYPE TRIPLET
```

and arrays may be of derived type as in the example

```
TYPE(TRIPLET), DIMENSION(10) :: T
```

A single structure (for example, T(2)) is always regarded as a scalar, but it may have a component (for example, T(2)%VERTEX) that is an array. Derived types may have components of other derived types.

An object that is not part of a bigger object always has a name (up to 31 alphanumeric characters) and is called a *named object*. Its subobjects have *designators* (for example, T(1:7) and T(1)%VERTEX) that consist of the name of the object followed by one or more qualifiers. Each successive qualifier specifies a part of the object specified by the name or designator that precedes it.

Because of these possibilities, the terms 'array' and 'variable' are now used with a more general meaning than in Fortran 77. The term 'array' is used for any object that is not scalar, including an array section or an array-valued component of a structure. The term 'variable' is used for any named object that is not specified to be a constant and for any part of such an object, including array elements, array sections, structure components, and substrings.

## 2.13  Pointers

In everyday language, nouns are often used in a way that makes their meaning precise only because of the context. 'The chairman said that...' will be understood precisely by the reader who knows that the context is the Fortran Committee and that its chairman is Jeanne Adams.

Similarly, in a computer program it can be very useful to be able to use a name that can be made to refer to different objects during execution of the program. One example is the multiplication of a vector by a sequence of square matrices. We might write code that calculates

$$y_i = \sum_{j=1}^{n} a_{ij} x_j, \, i = 1,2,...n$$

from the vector $x_j, j = 1,2,...n$. In order to use this to calculate

$$BCz$$

we might first make $x$ refer to $z$ and $A$ refer to $C$, thereby using our code to calculate $y=Cz$, then make $x$ refer to $y$ and $A$ refer to $B$ so that our code calculates the result vector we finally want.

An object that can be made to refer to other objects in this way is called a *pointer*, and must be declared with the pointer attribute, for example

```
REAL, POINTER            :: SON
REAL, POINTER, DIMENSION(:)   :: X, Y
REAL, POINTER, DIMENSION(:,:) :: A
```

In the case of an array, only the rank (number of dimensions) is declared, and the bounds (and hence shape) are taken from that of the object to which it points.

Besides pointing to existing variables, a pointer may be given fresh storage by an ALLOCATE statement such as

```
ALLOCATE (SON, X(10), Y(-10,10), A(N, N))
```

In the case of arrays, the lower and upper bounds are specified just as for the DIMENSION attribute (Section 2.10) except that any scalar integer expression is permitted. Permitting such expressions remedies one of the major deficiencies of Fortran 77, namely that areas of only static storage may be defined, and we will discuss this use further in Section 6.7.

Components of derived types are permitted to have the pointer attribute. This enables a major application of pointers: the construction of linked lists. As a simple example, we might decide to hold a sparse vector as a chain of variables of type shown in Figure 3, which allows us to access the entries one by one. Additional entries may be created when necessary by an appropriate ALLOCATE statement. We defer the details to Section 3.12.

---

```
TYPE ENTRY
   REAL VALUE
   INTEGER INDEX
   TYPE(ENTRY), POINTER :: NEXT
END TYPE ENTRY
```

---

Figure 3.

Apart from this case, subobjects are not pointers.

## 2.14 Summary

In this chapter we have introduced the elements of the Fortran 90 language. The character set has been listed, and the manner in which sequences of characters form literal constants and names explained. In this context we have encountered the five intrinsic data types defined in Fortran 90, and seen how each data type has corresponding literal constants and named objects. We have seen how derived types may be constructed from the intrinsic types. We have introduced one method by which arrays may be declared, and seen how their elements may be referenced by subscript expressions. The concepts of the array section, character substring, and pointer have been presented, and some important terms defined. In the following chapter we shall see how these elements may be combined into expressions and statements, Fortran's equivalents of 'phrases' and 'sentences'.

With respect to Fortran 77 there are many changes in this area: the larger character set; a new source form; the significance of blanks; the parameterization of the intrinsic types; the binary, octal, and hexadecimal constants; the ability to obtain a desired precision and range; quotes as well as apostrophes as character constant delimiters; longer names; derived data types; new array subscript notations; array constructors; and last, but not least, pointers. Together they represent a substantial improvement in the ease of use and power of the language.

## Exercises

**1.** For each of the following assertions, state whether it is true, false or not determined, according to the Fortran 90 collating sequences:

> B is less than M
> 8 is less than 2
> * is greater than T
> $ is less than /
> blank is greater than A
> blank is less than 6

**2.** Which of the Fortran 90 lines in Figure 4 are correctly written according to the requirements of the Fortran 90 source form? Which ones contain commentary? Which lines are initial lines and which are continuation lines?

```
     X = Y
 3   A = B+C ! Add
     WORD = 'String'
     A = 1.0; B = 2.0
     A = 15. ! Initialize A; B = 22. ! and B
     SONG = "Life is just&
        & a bowl of cherries"
     CHIDE = 'Waste not,
        want not!'
 0   C(3:4) = 'UP"
```

Figure 4.

**3.** Classify the following literal constants according to the five intrinsic data types of Fortran 90. Which are not legal literal constants?

```
-43                'WORD'
4.39               1.9-4
0.0001E+20         'STUFF & NONSENSE'
4 9                (0.,1.)
(1.E3,2)           'I CAN''T'
'(4.3E9, 6.2)'     .TRUE._1
E5                 'SHOULDN' 'T'
1_2                "O.K."
Z10                Z'10'
```

**4.** Which of the following names are legal Fortran 90 names?

```
NAME        NAME32
QUOTIENT    123
A182C3      NO-GO
STOP!       BURN_
NO_GO       LONG__NAME
```

**5.** What are the first, tenth, eleventh and last elements of the following arrays?

```
REAL, DIMENSION(11)      :: A
REAL, DIMENSION(0:11)    :: B
REAL, DIMENSION(-11:0)   :: C
REAL, DIMENSION(10,10)   :: D
REAL, DIMENSION(5,9)     :: E
REAL, DIMENSION(5,0:1,4) :: F
```

Write an array constructor of eleven integer elements.

**6.** Given the array declaration

```
CHARACTER(LEN=10), DIMENSION(0:5,3) :: C
```

which of the following subobject designators are legal?

```
C(2,3)          C(4:3)(2,1)
C(6,2)          C(5,3)(9:9)
C(0,3)          C(2,1)(4:8)
C(4,3)(:)       C(3,2)(0:9)
C(5)(2:3)       C(5:6)
C(5,3)(9)       C(,)
```

**7.** Write derived type definitions appropriate for:

    a) a vehicle registration;
    b) a circle;
    c) a book (title, author, and number of pages).

Give an example of a derived type constant for each one.

**8.** Given the declaration for T in Section 2.12, which of the following objects and subobjects are arrays?

```
T               T(4)%VERTEX(1)
T(10)           T(5:6)
T(1)%VERTEX     T(5:5)
```

**9.** Write specifications for these entities:

    a) an integer variable inside the range $-10^{20}$ to $10^{20}$;
    b) a real variable with a minimum of 12 decimal digits of precision and a
       range of $10^{-100}$ to $10^{100}$;
    c) a Kanji character variable on a processor that supports Kanji with
       KIND=2.

# 3. EXPRESSIONS AND ASSIGNMENTS

## 3.1 Introduction

We have seen in the previous chapter how we are able to build the 'words' of Fortran 90 — the constants, keywords, and names — from the basic elements of the character set. In this chapter we shall discover how these entities may be further combined into 'phrases' or *expressions,* and how these, in turn, may be combined into 'sentences', or *statements.*

In an expression, we describe a computation that is to be carried out by the computer. The result of the computation may then be assigned to a variable. A sequence of assignments is the way in which we specify, step-by-step, the series of individual computations to be carried out, in order to arrive at the desired result. There are separate sets of rules for expressions and assignments, depending on whether the operands in question are numeric, logical, character, or derived in type, and whether they are scalars or arrays. We shall discuss each set of rules in turn, including a description of the relational expressions which produce a result of type logical and are needed in control statements (see next chapter). To simplify the initial discussion, we commence by considering expressions and assignments that are intrinsically defined and involve neither arrays nor entities of derived data types.

An expression in Fortran 90 is formed of operands and operators, combined in a way which follows the rules of Fortran 90 syntax. A simple expression involving a *dyadic* (or binary) operator has the form

operand *operator* operand

an example being

X+Y

and a unary or *monadic* operator has the form

*operator* operand

an example being

−Y

The operands may be constants, variables, or functions (see Chapter 5), and an expression may itself be used as an operand. In this way we can build up more complicated expressions such as

operand *operator* operand *operator* operand

where consecutive operands are separated by a single operator. Each operand must have a defined value and the result must be mathematically defined; for example, dividing by zero is not permitted. Operators may be *intrinsic* (always available) or *defined* (see Section 3.8).

The rules of Fortran 90 state that the parts of expressions without parentheses are evaluated successively from left to right for operators of equal precedence, with the exception of ** (see Section 3.2). If it is necessary to evaluate part of an expression, or *subexpression*, before another, parentheses may be used to indicate which subexpression should be evaluated first. In

operand *operator* (operand *operator* operand)

the subexpression in parentheses will be evaluated first, and the result used as an operand to the first operator.

If an expression or subexpression has no parentheses, the processor is permitted to evaluate an equivalent expression, that is an expression that always has the same value apart, possibly, from the effects of numerical round-off. For example, if A, B, and C are real variables, the expression

A/B/C

might be evaluated as

A/(B*C)

on a processor that can multiply much faster than it can divide. Usually, such changes are welcome to the programmer since the program runs faster, but when they are not (for instance because they would lead to more round-off) parentheses should be inserted because the processor is required to respect them.

## 3.2  Scalar numeric expressions

A *numeric expression* is an expression whose operands are one of the three
numeric types — integer, real, and complex — and whose operators are

```
**          exponentiation
* /         multiplication, division
+ -         addition, subtraction
```

These operators are known as *numeric intrinsic* operators, and are listed here
in their order of precedence.  In the absence of parentheses, exponentiations
will be carried out before multiplications and divisions, and these before additions and subtractions.

We note that the minus sign (−) and the plus sign (+) can be used as a
unary operators, as in

```
−TAX
```

The type and kind type parameter of the result are those of the operand.

The exception to the left-to-right rule noted in Section 3.1 concerns
exponentiations.  Whereas the expression

```
−A+B+C
```

will be evaluated from left to right as

```
((−A)+B)+C
```

the expression

```
A**B**C
```

will be evaluated as

```
A**(B**C)
```

For integer data, the result of any division will be truncated towards zero,
that is to the integer value whose magnitude is equal to or just less than the
magnitude of the exact result.  Thus, the result of

```
6/3     is 2
8/3     is 2
−8/3    is −2
```

This fact must always be borne in mind whenever integer divisions are written. Similarly, the result of

2**3      is 8

whereas the result of

2**(−3)   is 1/(2**3)

which is zero.

The rules of Fortran 90 allow a numeric expression to contain numeric operands of differing types or kind type parameters. This is known as a *mixed-mode expression*. Except when raising a real or complex value to an integer power, the object of the weaker (or simpler) of the two data types will be converted, or *coerced*, into the type of the stronger one. The result will also be that of the stronger type. If, for example, we write

A*I

when A is of type real and I is of type integer, then I will be converted to a real data type before the multiplication is performed, and the result of the computation will also be of type real. The rules are summarized for each possible combination for the operations +, −, * and / in Table 2, and for the operation ** in Table 3. The functions REAL, CMPLX, and KIND that they reference are defined in Section 8.3.1. In both Tables, I stands for integer, R for real, and C for complex.

Table 2. Type of result of. *a* .op. *b*, where .op. is +, −, * or /.

| Type of *a* | Type of *b* | Value of *a* used | Value of *b* used | Type of result |
|---|---|---|---|---|
| I | I | *a* | *b* | I |
| I | R | REAL(*a*,KIND(*b*)) | *b* | R |
| I | C | CMPLX(*a*,KIND(*b*)) | *b* | C |
| R | I | *a* | REAL(*b*,KIND(*a*)) | R |
| R | R | *a* | *b* | R |
| R | C | CMPLX(*a*,KIND(*b*)) | *b* | C |
| C | I | *a* | CMPLX(*b*,KIND(*a*)) | C |
| C | R | *a* | CMPLX(*b*,KIND(*a*)) | C |
| C | C | *a* | *b* | C |

If both operands are of type integer, the kind type parameter of the result is that of the operand with the greater decimal exponent range, or is processor dependent if the kinds differ but the decimal exponent ranges are the same. If both operands are of type real or complex, the kind type parameter of the result is that of the operand with the greater decimal precision, or is processor dependent if the kinds differ but the decimal precisions are the same. If one operand is of type integer and the other is of real or complex, the type parameter of the result is that of the real or complex operand.

Table 3. Type of result of $a**b$.

| Type of $a$ | Type of $b$ | Value of $a$ used | Value of $b$ used | Type of result |
|---|---|---|---|---|
| I | I | $a$ | $b$ | I |
| I | R | REAL($a$,KIND($b$)) | $b$ | R |
| I | C | CMPLX($a$,KIND($b$)) | $b$ | C |
| R | I | $a$ | $b$ | R |
| R | R | $a$ | $b$ | R |
| R | C | CMPLX($a$,KIND($b$)) | $b$ | C |
| C | I | $a$ | $b$ | C |
| C | R | $a$ | CMPLX($b$,KIND($a$)) | C |
| C | C | $a$ | $b$ | C |

In the case of raising a complex value to a complex power, the principal value[3] is taken.

## 3.3  Defined and undefined variables

In the course of the explanations in this and the following chapters, we shall often refer to a variable becoming *defined* or *undefined*. In the previous chapter, we showed how a scalar variable may be called into existence by a statement like

```
REAL SPEED
```

In this simple case, the variable SPEED has, at the beginning of the execution of the program, no defined value. It is undefined. No attempt must be made to reference its value since it has none. A common way in which it might become defined is for it to be assigned a value:

```
SPEED = 2.997
```

---

[3]  The principal value of $a^b$ is $\exp(b(\log|a|+i\arg a))$, with $-\pi<\arg a\le\pi$.

After the execution of such an *assignment statement* it has a value, and that value may be referenced, for instance in an expression:

```
SPEED*0.5
```

For a compound object it is necessary for all its subobjects to be individually defined before the object as a whole is regarded as defined. Thus, an array is said to be defined only when each of its elements is defined, an object of a derived data type is defined only when each of its components is defined, and a character variable is defined only when each of its characters is defined.

A variable that is defined does not necessarily retain its state of definition throughout the execution of a program. As we shall see in Chapter 5, a variable that is local to a single subprogram usually becomes undefined when control is returned from that subprogram. In certain circumstances it is even possible that a single array element becomes undefined: this causes the array considered as a whole to become undefined; a similar rule holds for entities of derived data type and for character variables.

In the case of a pointer, the pointer association status may be *undefined* (its initial state), *associated* with a target, or *disassociated*, which means that it is not associated with a target but has a definite status that may be tested by the function ASSOCIATED (Section 8.2). Even though a pointer is associated with a target, the target itself may be defined or undefined.

## 3.4  Scalar numeric assignment

The general form of a scalar numeric assignment is

　　*variable* = *expr*

where *variable* is a scalar numeric variable and *expr* is a scalar numeric expression. If *expr* is not of the same type or kind as *variable*, it will be converted to that type and kind before the assignment is carried out, according to the set of rules given in Table 4.

Table 4. Numeric conversion for assignment statement  *variable* = *expr*

| Type of *variable* | Value assigned |
| --- | --- |
| integer | INT(*expr*, KIND(*variable*)) |
| real | REAL(*expr*, KIND(*variable*)) |
| complex | CMPLX(*expr*, KIND(*variable*)) |

We note that if the type of *variable* is integer but *expr* is not, then the assignment will result in a loss of precision unless *expr* happens to have an integral value. Similarly, assigning a real expression to a real variable of a kind with less precision will also cause a loss of precision to occur, and the assignment of a complex quantity to a non-complex variable involves the loss of the imaginary part. Thus, the values in I and A following the assignments

```
I = 7.3                     ! I of type default integer
A = (4.01935, 2.12372)      ! A of type default real
```

are 7 and 4.01935, respectively.

## 3.5  Scalar relational operators

It is possible in Fortran 90 to test whether the value of one numeric expression bears a certain relation to that of another, and similarly for character expressions. The relational operators are

```
.LT. or <        less than
.LE. or <=       less than or equal
.EQ. or ==       equal
.NE. or /=       not equal
.GT. or >        greater than
.GE. or >=       greater than or equal
```

If either or both of the expressions are complex, only the operators == and /= (or .EQ. and .NE.) are available.

The result of such a comparison is one of the default logical values .TRUE. or .FALSE., and we shall see in the next chapter how such tests are of great importance in controlling the flow of a program. Examples of relational expressions (for I and J of type integer, A and B of type real, and CHAR1 of type default character) are

```
I .LT. 0         integer relational expression
A < B            real relational expression
A+B .GT. I-J     mixed-mode relational expression
CHAR1 == 'Z'     character relational expression
```

In the third expression above, we note that the two components are of different numeric types. In this case, and whenever either or both of the two components consist of numeric expressions, the rules state that the components are to be evaluated separately, and converted to the type and kind of their sum before the comparison is made. Thus, a relational expression such as

```
A+B .LE. I-J
```

will be evaluated by converting the result of (I–J) to type real.

For character comparisons, the kinds must be the same and the letters are compared from the left until a difference is found or the strings are found to be identical. If the lengths differ, the shorter one is regarded as being padded with blanks[4] on the right. Two zero-sized strings are considered to be identical.

No other form of mixed mode relational operator is intrinsically available, though such an operator may be defined (Section 3.8). The numeric operators take precedence over the relational operators.

## 3.6  Scalar logical expressions and assignments

Logical constants, variables, and functions may appear as operands in logical expressions. The logical operators, in decreasing order of precedence, are:

*unary operator:*

```
.NOT.            logical negation
```

*binary operators:*

```
.AND.            logical intersection
.OR.             logical union
.EQV. and .NEQV. logical equivalence and non-equivalence
```

If we assume a logical declaration of the form

```
LOGICAL I,J,K,L
```

then the following are valid logical expressions:

```
.NOT.J
J .AND. K
I .OR. L .AND.  .NOT.J
( .NOT.K .AND. J .NEQV. .NOT.L) .OR. I
```

---

[4] Here and elsewhere, the blank padding character used for a nondefault type is processor dependent.

In the first expression we note the use of .NOT. as a unary operator. In the third expression, the rules of precedence imply that the subexpression L.AND..NOT.J will be evaluated first, and the result combined with I. In the last expression, the two subexpressions .NOT.K.AND.J and .NOT.L will be evaluated and compared for non-equivalence. The result of the comparison, .TRUE. or .FALSE., will be combined with I.

The kind type parameter of the result is that of the operand for .NOT., and for the others is that of the operands if they have the same kind or processor dependent otherwise.

We note that the .OR. operator is an inclusive operator; there is no exclusive logical OR (A.AND..NOT.B .OR. .NOT.A.AND.B).

The result of any logical expression is the value true or false, and this value may then be assigned to a logical variable such as element 3 of array FLAG in the example

```
FLAG(3) = ( .NOT. K .EQV. L) .OR. J
```

The kind type parameter values of the variable and expression need not be identical.

A logical variable may be set to a predetermined value by an assignment statement:

```
FLAG(1) = .TRUE.
FLAG(2) = .FALSE.
```

In the foregoing examples all the operands and results were of type logical — no other data type is allowed to participate in an intrinsic logical operation or assignment.

The results of several relational expressions may be combined into a logical expression, and assigned, as in

```
REAL A, B, X, Y
LOGICAL COND
:
COND = A>B .OR. X<0. .AND. Y>1.
```

where we note the precedence of the relational operators over the logical operators.

## 3.7  Scalar character expressions and assignments

The only intrinsic operator for character expressions is the concatenation operator //, which has the effect of combining two character operands into a single character result.  For example, the result of concatenating the two character constants 'AB' and 'CD', written as

```
'AB'//'CD'
```

is the character string ABCD.  The operands must have the same kind parameter values, but may be character variables, constants, or functions.  For instance, if WORD1 and WORD2 are both of default kind and length 4, and contain the character strings LOOP and HOLE respectively, the result of

```
WORD1(4:4)//WORD2(2:4)
```

is the string POLE.

The length of the result of a concatenation is the sum of the lengths of the operands.  Thus, the length of the result of

```
WORD1//WORD2//'S'
```

is 9, which is the length of the string LOOPHOLES.

The result of a character expression may be assigned to a character variable of the same kind.  Assuming the declarations

```
CHARACTER(LEN=4) CHAR1, CHAR2
CHARACTER(LEN=8) RESULT
```

we may write

```
CHAR1 = 'ANY '
CHAR2 = 'BOOK'
RESULT = CHAR1//CHAR2
```

In this case, RESULT will now contain the string ANY BOOK.  We note in these examples that the lengths of the left- and right-hand sides of the three assignments are in each case equal.  If, however, the length of the result of the right-hand side is shorter than the length of the left-hand side, then the result is placed in the left-most part of the left-hand side and the rest is filled with blank characters.  Thus, in

```
CHARACTER(LEN=5) FILL
FILL(1:4) = 'AB'
```

FILL(1:4) will have the value AB*bb* (where *b* stands for a blank character). The value of FILL(5:5) remains undefined, that is, it contains no specific value and should not be used in an expression. As a consequence, FILL is also undefined. On the other hand, when the left-hand side is shorter than the result of the right-hand side, the right-hand end of the result is truncated. The result of

```
CHARACTER(LEN=5) TRUNC8
TRUNC8 = 'TRUNCATE'
```

is to place in TRUNC8 the character string TRUNC. If a left-hand side is of zero length, no assignment takes place.

The left-hand and right-hand sides of an assignment may overlap. In such a case, it is always the old values that are used in the right-hand side expression. For example, the assignment

```
RESULT(3:5) = RESULT(1:3)
```

is valid and if RESULT began with the value ABCDEFGH, it would be left with the value ABABCFGH.

## 3.8  Structure constructors and scalar defined operators

No operators for derived types are automatically available, but a structure may be constructed from expressions for its components, just as a constant structure may be constructed from constants (Section 2.9). The general form of a *structure constructor* is

   *type-name (expr-list)*

where the *expr-list* specifies the values of the components. For example, given the type

```
TYPE STRING
   INTEGER LENGTH
   CHARACTER(LEN=10) VALUE
END TYPE STRING
```

and the variables

```
CHARACTER(LEN=4) CHAR1, CHAR2
```

the following is a value of type STRING

```
STRING(8, CHAR1//CHAR2)
```

Each expression in *expr-list* must agree in rank with the corresponding component of the structure; if it is not a pointer component, it must agree in shape and its value is assigned to the component under the rules of intrinsic assignment; if it is a pointer component, the expression must be a valid target for it,[5] as in a pointer assignment statement (Section 3.12).

When a programmer defines a derived type and wishes operators to be available, he or she must define the operators, too. For a binary operator this is done by writing a function, with two arguments, that specifies how the result depends on the operands, and an interface block that associates the function with the operator token (functions and interface blocks will be explained fully in Chapter 5). For example, given the type

```
TYPE INTERVAL
   REAL LOWER, UPPER
END TYPE INTERVAL
```

that represents intervals of numbers between a lower and an upper bound, we may define addition by a module (Section 5.5) containing the procedure

```
FUNCTION ADD_INTERVALS(A,B)
   TYPE(INTERVAL) ADD_INTERVALS, A, B
   ADD_INTERVALS%LOWER = A%LOWER + B%LOWER
   ADD_INTERVALS%UPPER = A%UPPER + B%UPPER
END FUNCTION ADD_INTERVALS
```

and the interface block (Section 5.18)

```
INTERFACE OPERATOR(+)
   MODULE PROCEDURE ADD_INTERVALS
END INTERFACE
```

This function would be invoked in an expression such as

```
X = Y + Z
```

---

5   In particular, it must not be a constant.

to perform this programmer-defined add operation for scalar variables X, Y, and Z of type INTERVAL. A unary operator is defined by an inteface block and a function with one argument.

The operator token may be any of the tokens used for the intrinsic operators or may be a sequence of up to 31 letters enclosed in decimal points, such as

```
.SUM.
```

In this case, the header line of the interface block would be written as

```
INTERFACE OPERATOR(.SUM.)
```

and the expression as

```
X = Y.SUM.Z
```

If an intrinsic token is used, the number of arguments must be the same as for the intrinsic operation and the priority of the operation is as for the intrinsic operation. Otherwise, it is of highest priority for defined unary operators and lowest priority for defined binary operators. The complete set of priorities is given in Table 5. Where another priority is required within an expression, parentheses must be used.

Table 5. Relative precedence of operators (in decreasing order).

| Type of operation when intrinsic | Operator |
|---|---|
| — | monadic (unary) operator |
| Numeric | ** |
| Numeric | * or / |
| Numeric | monadic + or − |
| Numeric | dyadic + or − |
| Character | // |
| Relational | .EQ.  .NE.  .LT.  .LE.  .GT.  .GE. |
| | ==   /=   <   <=   >   >= |
| Logical | .NOT. |
| Logical | .AND. |
| Logical | .OR. |
| Logical | .EQV. or .NEQV. |
| — | dyadic (binary) operator |

Retaining the intrinsic priorities is helpful both to the readability of expressions and to the efficiency with which a compiler can interpret them. For example, if + is used for set union and * for set intersection, we can interpret the expression

```
I*J + K
```

for sets I, J, and K without difficulty.

If either of the intrinsic tokens .EQ. and == is used, the definition applies to both tokens so that they are always equivalent. The same is true for the other equivalent pairs of relational operators.

Although no two intrinsic operators may follow one another, a binary operator or an intrinsic unary operator may be followed by a defined unary operator.

Operators may be defined for any types of operands, except where there is an intrinsic operation for the operator and types. For example, we might wish to be able to add an interval number to an ordinary real, which can be done by adding the procedure

```
FUNCTION ADD_INTERVAL_REAL(A,B)
    TYPE(INTERVAL) ADD_INTERVAL_REAL, A
    REAL B
    ADD_INTERVAL_REAL%LOWER = A%LOWER + B
    ADD_INTERVAL_REAL%UPPER = A%UPPER + B
END FUNCTION ADD_INTERVAL_REAL
```

and changing the interface block to

```
INTERFACE OPERATOR(+)
    MODULE PROCEDURE ADD_INTERVALS, ADD_INTERVAL_REAL
END INTERFACE
```

The result of a defined operation may have any type. The type of the result, as well as its value, must be specified by the function. Note that an operation that is defined intrinsically cannot be redefined; thus in

```
REAL A, B, C
C = A + B
```

the meaning of the operation is always unambiguous.

## 3.9  Scalar defined assignments

Assignment of an expression of derived type to a variable of the same type is automatically available.  For example, if A is of the type INTERVAL defined at the start of Section 3.8, we may write

```
A = INTERVAL(0., 1.)
```

(structure constructors were met in Section 3.8, too).  This assignment may be redefined or another assignment may be defined by a subroutine with two arguments, the first corresponding to the variable and the second to the expression (subroutines will also be dealt with fully in Chapter 5).  For example, assignment of reals to intervals and vice-versa might be defined by a module containing the subroutines

```
SUBROUTINE REAL_FROM_INTERVAL(A,B)
   REAL A
   TYPE(INTERVAL) B
   A = (B%LOWER + B%UPPER)/2
END SUBROUTINE
```

and

```
SUBROUTINE INTERVAL_FROM_REAL(A,B)
   TYPE(INTERVAL) A
   REAL B
   A%LOWER = B
   A%UPPER = B
END SUBROUTINE
```

and the interface block

```
INTERFACE ASSIGNMENT(=)
   MODULE PROCEDURE REAL_FROM_INTERVAL, INTERVAL_FROM_REAL
END INTERFACE
```

Given this we may write

```
TYPE(INTERVAL) A
A = 0.
```

A defined assignment may not redefine the meaning of an intrinsic assignment for intrinsic types, that is an assignment between two objects of numeric type, logical type, or character type with the same kind parameter, but may redefine the meaning of an intrinsic assignment for two objects of the same derived type.

## 3.10 Array expressions

So far in this chapter, we have assumed that all the entities in an expression are scalar. However, any of the unary intrinsic operations may also be applied to an array to produce another array of the same shape (identical rank and extents, see Section 2.10) and having each element value equal to that of the operation applied to the corresponding element of the operand. Similarly, binary intrinsic operations may be applied to a pair of arrays of the same shape to produce an array of that shape, with each element value equal to that of the operation applied to corresponding elements of the operands. One of the operands to a binary operation may be a scalar, in which case the result is as if the scalar had been broadcast to an array of the same shape as the array operand. Given the array declarations

```
REAL, DIMENSION(10,20) :: A,B
REAL, DIMENSION(5) :: V
```

the following are examples of array expressions:

```
A/B          ! Array of shape (10,20), with elements A(I,J)/B(I,J)
V+1.         ! Array of shape (5), with elements V(I)+1.0
5/V+A(1:5,5) ! Array of shape (5), with elements 5/V(I)+A(I,5)
A.EQ.B       ! Logical array of shape (10,20), with elements
             ! .TRUE. if A(I,J).EQ.B(I,J), and .FALSE. otherwise
```

Two arrays of the same shape are said to be *conformable* and a scalar is conformable with any array.

Note that the correspondence is by position in the extent and not by subscript value. For example,

```
A(2:9,5:10) + B(1:8,15:20)
```

has element values

```
A(I+1,J+4) + B(I,J+14), I=1,2,...,8, J=1,2,...,6.
```

This may be represented pictorially as in Figure 5.

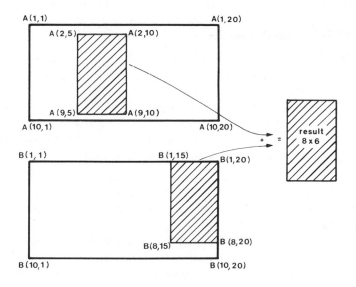

Figure 5. The sum of two array sections.

The order in which the scalar operations in any array expression are executed is not specified in the standard, thus enabling a compiler to arrange efficient execution on a vector or parallel computer.

Any scalar intrinsic operator may be applied in this way to arrays and array-scalar pairs. For derived operators, the programmer must define operators directly for array operands, for each rank or pair of ranks involved. For example, the type

```
TYPE MATRIX
    REAL ELEMENT
END TYPE MATRIX
```

might be defined to have scalar operations that are identical to the operations for reals, but for arrays of ranks one and two the operator * defined to mean matrix multiplication. The type MATRIX would therefore be suitable for matrix arithmetic, whereas reals are not suitable because multiplication for real arrays is done element by element. This is further discussed in Section 6.6.

## 3.11  Array assignment

An array expression may be assigned to an array variable of the same shape, which is interpreted as if each element of the expression were assigned to the corresponding element of the variable.  For example, with the declarations of the beginning of the last section, the assignment

```
A = A + 1.0
```

replaces A(I,J) by A(I,J) + 1.0 for I = 1, 2,..., 10 and J = 1, 2,..., 20.  Note that, as for expressions, the element correspondence is by position within the extent rather than by subscript value.  This is illustrated by the example

```
A(1,11:15) = V     ! A(1,J+10) is assigned from
                   ! V(J), J=1,2,...,5
```

A scalar intrinsic expression may be assigned to an intrinsic array, in which case the scalar value is broadcast to all the array elements.

If the expression includes a reference to the array variable or to a part of it, the expression is interpreted as being fully evaluated before the assignment commences.  For example, the statement

```
V(2:5) = V(1:4)
```

results in each element V(I), I = 2, 3, 4, 5 having the value that V(I–1) had prior to the commencement of the assignment.  This rule exactly parallels the rule for substrings that was explained in Section 3.7.  The order in which the array elements are assigned is not specified by the standard, in order to allow optimizations.

## 3.12  Pointers in expressions and assignments

A pointer may appear as a variable in the expressions and assignments that we have considered so far in this chapter, provided it has a valid association with a target.  The target is accessed implicitly (that is, without any need for an explicit dereferencing symbol).  In particular, if two pointers appear on opposite sides of an assignment statement, data are copied from one target to the other target.

Sometimes the need arises for another sort of assignment.  We may want the left-hand pointer to point to another target, rather than that its current target acquire fresh data.  This is called *pointer assignment* and takes place in a pointer assignment statement:

*pointer* => *target*

where *pointer* is the name of a pointer or the designator of a structure component that is a pointer, and *target* is usually a variable but may also be an
expression that has the POINTER attribute. For example, the statements

```
X => Z
A => C
```

are needed for the first matrix multiplication of Section 2.13, in order to make
X refer to Z and A to refer to C. Pointer assignment also takes place for a
pointer component of a structure when the structure appears on the left-hand
side of an ordinary assignment. For example, suppose we have used the type
ENTRY of Section 2.13 to construct a chain of entries and wish to add a fresh
entry at the front. If FIRST points to the first entry and the pointer variable
CURRENT is currently allocated, the statement

```
CURRENT = ENTRY(NEW_VALUE, NEW_INDEX, FIRST)
```

is suitable. It sets the pointer CURRENT%NEXT to the pointer FIRST rather
than altering its present target, and this is just what we need.

If the *target* in a pointer assignment statement is a variable that is not itself
a pointer or a subobject of a pointer, it must have the TARGET attribute. For
example, the statement

```
REAL, DIMENSION(10), TARGET :: Y
```

declares Y to have the TARGET attribute. Any subobject of an object with
the TARGET attribute also has the TARGET attribute. The TARGET attribute is required for the purpose of code optimization by the compiler. It is
very helpful to the compiler to know that a variable that is not a pointer or a
target may not be accessed by a pointer.

The target in a pointer assignment statement may be a subobject of a
pointer that is selected without further component selection. For example,
given the declaration

```
CHARACTER(LEN=80), POINTER :: PAGE(60)
```

the following are all permitted targets:

```
PAGE, PAGE(10), PAGE(2:4), PAGE(2)(3:15)
```

If the *target* in a pointer assignment statement is itself a pointer, then a straightforward copy of the pointer takes place. If it is undefined or disassociated, this state is copied; otherwise the targets become identical.

The type, type parameters, and rank of the *pointer* and *target* in a pointer assignment statement must each be the same. If the *pointer* is an array, it takes its shape and bounds from the *target*. The bounds are as would be returned by the functions LBOUND and UBOUND (Section 8.12.2) for the target, which means that an array section or array expression is always taken to have the value 1 for a lower bound and the extent for the corresponding upper bound.

## 3.13  Summary

In this chapter, we have seen how scalar and array expressions, of numeric, logical, character, and derived types may be formed; and how the corresponding assignments of the results may be made. The relational expressions and the use of pointers have also been presented. We now have the information required to write short sections of code forming a sequence of statements to be performed one after the other. In the following chapter we shall see how more complicated sequences, involving branching and iteration, may be built up.

Features described in this chapter which are new to Fortran are the use of the alternative representations <, <=,... for the relational operators; the ability of the two sides of a character assignment to overlap; structure constructors; defined operators and assignment; array expressions and assignment; and the use of pointers in expressions and assignment.

## Exercises

**1.** If all the variables are numeric scalars, which of the following are valid numeric expressions?

```
A+B              -C
A+-C             D+(-F)
(A+C)**(P+Q)     (A+C)(P+Q)
-(X+Y)**I        4.((A-D)-(A+4.*X)+1)
```

**2.** In the following expressions, add the parentheses which correspond to Fortran 90's rules of precedence (assuming A, C—F are real scalars, I—N are logical scalars, and B is a logical array), for example

```
A+D**2/C      becomes      A+((D**2)/C)

C+4.*F
4.*G-A+D/2.
A**E**C**D
A*E-C**D/A+E
I .AND. J .OR. K
.NOT. L .OR  .NOT. I .AND. M .NEQV. N
B(3).AND.B(1).OR.B(6).OR..NOT.B(2)
```

**3.** What are the results of the following expressions?

```
3+4/2       6/4/2
3.*4**2     3.**3/2
-1.**2      (-1.)**3
```

**4.** A scalar character variable R has length 8. What are the contents of R after each of the following assignments?

```
R = 'ABCDEFGH'
R = 'ABCD'//'01234'
R(:7) = 'ABCDEFGH'
R(:6) = 'ABCD'
```

**5.** Which of the following logical expressions are valid, (B is a logical array)?

```
.NOT.B(1).AND.B(2)     .OR.B(1)
B(1).OR..NOT.B(4)      B(2)(.AND.B(3).OR.B(4))
```

**6.** If all the variables are real scalars, which of the following relational expressions are valid?

```
D .LE. C           P .LT. T > 0.
X-1 /= Y           X+Y < 3 .OR. > 4.
D.LT.C.AND.3.0     Q.EQ.R .AND. S>T
```

**7.** Write expressions to compute:

   a) the perimeter of a square of side L;

   b) the area of a triangle of base B and height H;

   c) the volume of a sphere of radius R.

**8.** An item costs $n$ cents. Write a declaration statement for suitable variables and assignment statements which compute the change to be given from a $1 bill for any value of $n$ from 1 to 99, using coins of denomination 1, 5, 10, and 25 cents.

**9.** Given the type declaration for INTERVAL in Section 3.8, the definitions of + given in Section 3.8, the definitions of assignment given in Section 3.9, and the declarations

```
TYPE(INTERVAL) A,B,C,D
REAL R
```

• which of the following statements are valid?

```
A = B + C
C = B + 1.0
D = B + 1
R = B + C
A = R + 2
```

**10.** Given the type declarations

```
REAL, DIMENSION(5,6) :: A, B
REAL, DIMENSION(5) :: C
```

which of the following statements are valid?

```
A = B            C = A(:,2) + B(5,:5)
A = C+1.0        C = A(2,:) + B(:,5)
A(:,3) = C       B(2:,3) = C + B(:5,3)
```

# 4. CONTROL STATEMENTS

## 4.1 Introduction

We have learnt in the previous chapter how assignment statements may be written, and how these may be ordered one after the other to form a sequence of code which is executed step-by-step. In most computations, however, this simple sequence of statements is by itself inadequate for the formulation of the problem. For instance, we may wish to follow one of two possible paths through a section of code, depending on whether a calculated value is positive or negative. We may wish to sum 1000 elements of an array, and to do this by writing 1000 additions and assignments is clearly tedious; the ability to iterate over a single addition is required instead. We may wish to pass control from one part of a program to another, or even stop processing altogether.

For all these purposes we have available in Fortran 90 various facilities to enable the logical flow through the program statements to be controlled. The facilities contained in Fortran 90 correspond to those now widely regarded as being the most appropriate for a modern programming language. Their general form is that of a *block* construct, that is a construct which begins with an initial keyword statement, may have intermediate keyword statements, and ends with a matching terminal statement, and that may be entered only at the initial statement. Each sequence of statements between keywords is called a *block*. A block may be empty, though such cases are rare.

Executable constructs may be *nested,* that is a block may contain another executable construct. In such a case, the block must contain the whole of the inner construct. Execution of a block always begins with its first executable statement.

We begin by describing the GO TO statement.

## 4.2 GO TO statement

In this section we consider the most disputed statement in programming languages — the GO TO statement. It is generally accepted that it is difficult to understand a program which is interrupted by many branches, especially if there is a large number of backward branches — those returning control to a statement preceding the branch itself. At the same time there are certain occasions, especially when dealing with error conditions, when GO TO statements are required in even the most advanced languages.

The form of the unconditional GO TO is

GO TO *label*

where *label* is a statement label. This statement label must be present on an *executable statement* (a statement which can be executed, as opposed to one of an informative nature, like a declaration). An example is

```
    X = Y+3.
    GO TO 4
 3  X = X+2.
 4  Z = X+Y
```

in which we note that after execution of the first statement, a branch is taken to the last statement, labelled 4. This is a *branch target statement*. The statement labelled 3 is jumped over, and can be executed only if there is a branch to the label 3 somewhere else. If the statement following an unconditional GO TO is unlabelled it can never be reached and executed, creating *dead code,* normally a sign of incorrect coding.

A GO TO statement must never specify a branch into a block, though it may specify a branch

- from within a block to another statement in the block,

- to the terminal statement of its construct, or

- to a statement outside its construct.

## 4.3  IF statement and construct

The IF statements provide a mechanism for branching depending on a condition. There are two types of constructs, known as the IF statement and the IF construct. They are powerful tools, the IF construct being a generalized form of the IF statement.

### 4.3.1  IF statement

In the IF statement, the value of a scalar logical expression is tested, and a single statement executed if its value is true. The general form is

IF (*scalar-logical-expr*) *action-stmt*

where *scalar-logical-expr* is any scalar logical expression, and *action-stmt* is any executable statement other than one that marks the beginning or end of a block (for instance, IF, ELSE IF, ELSE, END IF, see next subsection), another IF statement, or an END statement (see Chapter 5). Examples are

```
IF (FLAG) GO TO 6
IF (X-Y > 0.) X = 0.
IF (COND .OR. P.LT.Q .AND. R.LE.1.) S(I,J) = T(J,I)
```

The IF statement is normally used either to perform a single assignment depending on a condition, or to branch depending on a condition. The *action-stmt* may not be labelled separately.

## 4.3.2  IF construct

The IF construct allows either the execution of a sequence of statements (a block) to depend on a condition, or the execution of alternative sequences of statements (blocks) to depend on alternative conditions. The simplest of its three forms is

[ *name*:]  IF (*scalar-logical-expr*) THEN
    *block*
  END IF [*name*]

where *scalar-logical-expr* is any scalar logical expression and *block* is any sequence of executable statements (except an END statement or an incomplete construct). The *block* is executed if *scalar-logical-expr* evaluates to the value true, and is not executed if it evaluates to the value false. The IF construct may be optionally named: the first and last statements may bear the same name, which may be any valid and distinct Fortran name (see Section 5.15 for a discussion on the scope of names). The fact that the name is optional is indicated here by the square brackets, a convention that will be followed throughout the book.

We notice that the IF construct is a compound statement, the beginning being marked by the IF...THEN, and the end by the END IF. An example is

```
SWAP: IF (X < Y) THEN
         TEMP = X
         X = Y
         Y = TEMP
      END IF SWAP
```

in which we notice also that the block inside the IF construct is indented with respect to its beginning and end. This is not obligatory, but makes the logic easier to understand, especially in nested IF constructs as we shall see at the end of this section.

In the second form of the IF construct, an alternative block of statements is executable, for the case where the condition is false. The general form is

[*name*:]  IF (*scalar-logical-expr*) THEN
      *block1*
     ELSE [*name*]
      *block2*
     END  IF [*name*]

in which the first block of statements (*block1*) is executed if the condition is true and the second block (*block2*), following the ELSE statement, is executed if the condition is false.  An example is

```
IF (X.LT.Y) THEN
   X = -X
ELSE
   Y = -Y
END IF
```

in which the sign of X is changed if X is less than Y, and the sign of Y is changed if X is greater than or equal to Y.

The third and most general type of IF construct uses the ELSE IF statement to make a series of independent tests, each of which has its associated block of statements.  The tests are made one after the other until one is fulfilled, and the associated statements of the relevant IF or ELSE IF block are executed.  Control then passes to the end of the IF construct.  If no test is fulfilled, no block is executed, unless there is a final 'catch-all' ELSE clause.  The general form is

[ *name*:]   IF (*scalar-logical-expr*) THEN
      *block*
     [ELSE  IF (*scalar-logical-expr*) THEN [*name*]
      *block*]...
     [ELSE [*name*]
      *block*]
     END  IF [*name*]

Here, and later in the book, we use the notation [ ]... to indicate an optional item that may occur any number of times (including zero).  There can be any number (including zero) of ELSE IF statements, and zero or one ELSE statements.  An ELSE or ELSE IF statement may be named only if the corresponding IF and END IF statements are named, and must be given the same name.

The statements within an IF construct may be labelled, but the labels must never be referenced in such a fashion as to pass control into the range of an IF construct from outside it, to an ELSE IF or ELSE statement, or into a block

of the construct from outside the block.  For example, the following IF construct is illegal:

```
        IF (TEMP.GT.100.) THEN
            GO TO 1                     ! illegal branch
            BOIL = .TRUE.
            STEAM = .TRUE.
        ELSE
    1       BOIL = .FALSE.
            LIQUID = .TRUE.
        END IF
```

It is permitted to pass control to an END IF statement from within its construct.  Execution of an END IF statement has no effect.

It is permitted to nest IF constructs within one another to an arbitrary
depth, as shown to two levels in Figure 6, in which we see again the necessity
to indent the code in order to be able to understand the logic easily.  For even
deeper nesting, naming is to be recommended.  The constructs must be properly nested, that is each construct must be wholly contained in a block of the
next outer construct.

```
        IF (I < 0) THEN
            IF (J < 0) THEN
                X = 0.
                Y = 0.
            ELSE
                Z = 0.
            END IF
        ELSE IF (K < 0) THEN
            Z = 1.
        ELSE
            X = 1.
            Y = 1.
        END IF
```

Figure 6.

## 4.4 CASE construct

Fortran 90 provides another means of selecting one of several options, rather
similar to that of the IF construct.  The principal differences between the two
constructs are that, for the CASE construct, only *one* expression is evaluated
for testing, and the evaluated expression may belong to no more than one of a
series of pre-defined sets of values.  The form of the CASE construct is shown
by:

```
[ name:]  SELECT CASE (expr)
          [CASE selector [name]
              block]...
          END SELECT [name]
```

As for the IF construct, the leading and trailing statements must either both be unnamed or both bear the same name; an intermediate statement may be named only if the leading statement is named and bears the same name. The expression *expr* must be scalar and of type character, logical, or integer, and the specified values in each *selector* must be of this type. In the character case, the lengths are permitted to differ, but not the kinds. In the logical and integer cases, the kinds may differ. The simplest form of *selector* is a scalar initialization expression[6] in parentheses, such as in the statement

```
CASE(1)
```

For character or integer *expr*, a range may be specified by an upper and a lower scalar initialization expression separated by a colon:

CASE (*low:high*)

Either *low* or *high*, but not both, may be absent; this is equivalent to specifying that the case is selected whenever *expr* evaluates to a value that is less than or equal to *high*, or greater than or equal to *low*, respectively. An example is shown in Figure 7.

```
SELECT CASE (NUMBER)     ! NUMBER of type integer
CASE (:-1)               ! all values below 0
   N_SIGN = -1
CASE (0)                 ! only 0
   N_SIGN = 0
CASE (1:)                ! all values above 0
   N_SIGN = 1
END SELECT
```

Figure 7.

---

[6] An initialization expression is a restricted form of constant expression (the restrictions being chosen for ease of implementation). The details are tedious and are deferred to Section 7.4. In this section, all examples are the simplest form of initialization expression: the literal constant.

The general form of *selector* is a list of non-overlapping values and ranges, all of the same type as *expr,* enclosed in parentheses, such as

```
CASE (1, 2, 7, 10:17, 23)
```

The form

```
CASE DEFAULT
```

is equivalent to a list of all the possible values of *expr* that are not included in the other selectors of the construct. Though we recommend that the values be in order, as in this example, this is not required. Overlapping values are not permitted within one *selector*, nor between different ones in the same construct.

There may be only a single DEFAULT *selector* in a given CASE construct, as shown in Figure 8. The CASE DEFAULT clause does not necessarily have to be the last clause of the CASE construct.

---

```
SELECT CASE (CH)          ! CH of type character
CASE ('C', 'D', 'R':)
   CH_TYPE = .TRUE.
CASE ('I':'N')
   INT_TYPE = .TRUE.
CASE DEFAULT
   REAL_TYPE = .TRUE.
END SELECT
```

---

Figure 8.

Since the selectors are not permitted to overlap, at most one may be satisfied; if none is satisfied, control passes to the next executable statement following the END SELECT statement.

Like the IF construct, CASE constructs may be nested inside one another. Branching to a statement in a case block is permitted only from another statement in the block, it is not permitted to branch to a CASE statement, and any branch to an END SELECT statement must be from within the CASE construct which it terminates.

## 4.5 DO construct

Many problems in mathematics require, for their representation in a programming language, the ability to *iterate*. If we wish to sum the elements of an array A of length 10, we could write

```
SUM = A(1)
SUM = SUM+A(2)
:
SUM = SUM+A(10)
```

which is clearly laborious. Fortran 90 provides a facility known as the DO construct which allows us to reduce these ten lines of code to

```
SUM = 0.
DO  I = 1,10
    SUM = SUM+A(I)
END DO
```

In this fragment of code we first set SUM to zero, and then require that the statement between the DO statement and the END DO statement shall be executed ten times. For each iteration there is an associated value of an index, kept in I, which assumes the value 1 for the first iteration through the loop, 2 for the second, and so on up to 10. I is a normal integer variable, but is subject to the rule that it must not be explicitly modified within the DO construct.

The DO statement has more general forms. If we wished to sum the fourth to ninth elements we would write

```
DO  I = 4, 9
```

thereby specifying the required first and last values of I. If, alternatively, we wished to sum all the odd elements, we would write

```
DO  I = 1, 9, 2
```

where the third of the three loop *parameters,* namely the 2, specifies that I is to be incremented in steps of 2, rather than by the default value of 1, which is assumed if no third parameter is given. In fact, we can go further still, as the parameters need not be constants at all, but integer expressions, as in

```
DO  I = J+4, M, -K(J)**2
```

in which the first value of I is J+4, and subsequent values are decremented by K(J)**2 until the value of M is reached.  DO constructs may thus run 'backwards' as well as 'forwards'.  If any of the three parameters is a variable or is an expression that involves a variable, the value of the variable may be modified within the loop without affecting the number of iterations, as the *initial* values of the parameters are used for the control of the loop.

The general form of this type of DO construct control clause is

[*name:*]  DO *variable* = *expr1, expr2* [*,expr3*]
     *block*
    END DO [*name*]

where *variable* is a named scalar integer variable, *expr1, expr2,* and *expr3* (*expr3* is optional but must be nonzero when present) are any valid scalar integer expressions, and *name* is the optional construct name.  The DO and END DO statements must either both bear the same *name*, or both be unnamed.

The number of iterations of a DO construct is given by the formula

$$\text{MAX}((\; expr2 - expr1 + expr3)/expr3,\; 0)$$

where MAX is a function which we shall meet in Section 8.3.2 and which returns either the value of the expression or zero, whichever is the larger. There is a consequence following from this definition, namely that if a loop begins with the statement

```
DO  I = 1, N
```

then its body will not be executed at all if the value of N on entering the loop is zero or less.  This is an example of the *zero-trip loop,* and results from the application of the MAX function.

A very simple form of the DO statement is

[*name:*] DO

which specifies an endless loop.  In practice, a means to exit from an endless loop, or indeed from any loop, is required, and this is provided in the form of the EXIT statement:

EXIT [ *name*]

where *name* is optional and is used to specify from which construct the exit should be taken in the case of nested constructs.  Execution of an EXIT state-

ment causes control to be transferred to the next executable statement after the
END DO statement to which it refers. If no name is specified, it terminates
execution of the innermost DO construct in which it is enclosed. As an
example of this form of the DO, suppose we have used the type ENTRY of
Section 2.13 to construct a chain of entries in a sparse vector, and we wish to
find the entry with index 10, known to be present. If FIRST points to the first
entry, the code in Figure 9 is suitable.

```
TYPE(ENTRY), POINTER :: FIRST, CURRENT
:
CURRENT => FIRST
DO
    IF (CURRENT%INDEX == 10) EXIT
    CURRENT => CURRENT%NEXT
END DO
```

Figure 9.

A related statement is the CYCLE statement

CYCLE [ *name*]

which transfers control to the END DO statement of the corresponding construct. Thus, if further iterations are still to be carried out, the next one is
initiated.

The value of a DO construct index (if present) is incremented at the end of
every loop iteration for use in the subsequent iteration. As the value of this
index is available outside the loop after its execution, we have three possible
situations, each illustrated by the following loop:

```
DO  I = 1, N
    :
    IF (I.EQ.J) EXIT
    :
END DO
L = I
```

If, at execution time, N has the value zero or less, I is set to 1 but the loop is
not executed, and control passes to the statement following the END DO
statement. If, on the other hand, N has a value which is greater than or equal
to J, an exit will be taken at the IF statement, and L will acquire the last value
of I, which is of course J. If, as a third possibility, the value of N is greater
than zero but less than J, the loop will be executed N times, with the successive values of I being 1, 2,... *etc.* up to N. When reaching the end of the
loop for the N*th* time, I will be incremented a final time, acquiring the value

N+1, which will then be assigned to L.  We see how important it is to make careful use of loop indices outside the DO block, especially when there is the possibility of the number of iterations taking on the boundary value of the maximum for the loop.

The DO block, just mentioned, is the sequence of statements between the DO statement and the END DO statement.  From anywhere outside a DO block, it is prohibited to jump into the block or to its END DO statement. The following sequence is thus illegal:

```
   GO TO 2          ! illegal branch
   :
   DO I = 1, N
      :
 2    A = B + C
      :
   END DO
```

It is similarly illegal for the range of a DO construct (or an IF, CASE, or WHERE construct, see Section 6.8), to be only partially contained in a block of another construct.  The construct must be completely contained in the block.  The following two sequences are thus legal:

```
   IF ( scalar-logical-expr) THEN
      DO I = 1, N
         :
      END DO
   ELSE
      :
   END IF
```

and

```
   DO I = 1, N
      IF (scalar-logical-expr) THEN
         :
      END IF
   END DO
```

but this third sequence is not:

```
IF (scalar-logical-expr) THEN
   DO I = 1, 10
      :
   END IF  ! illegal position of IF construct termination
   :
END DO
```

DO constructs may be nested provided that the range of one loop is completely contained within the range of another. We may thus write a matrix multiplication as shown in Figure 10.

```
DO  I = 1, N
   DO  J = 1, M
      A(I,J) = 0.
      DO  L = 1, K
         A(I,J) = A(I,J)+B(I,L)*C(L,J)
      END DO
   END DO
END DO
```

Figure 10.

Another example is the summation loop in Figure 11.

```
DO  I = 1, N
   SUM = 0.
   DO  J = 1,I
      SUM = SUM+B(J,I)
   END DO
   A(I) = SUM
END DO
```

Figure 11.

A final form of the DO construct makes use of a statement label to identify the end of the construct. In this case, the terminating statement may be either a labelled END DO statement or a labelled CONTINUE ('do nothing') statement.[7] The label is, in each case, the same as that on the DO statement itself. Simple examples are

---

[7] The CONTINUE statement is not limited to being the last statement of a DO construct; it may appear anywhere among the executable statements.

```
    DO 10 I = 1, N
       :
10 END DO
```

and

```
    DO 20 I = 1, J
       DO 10 K = 1, L
          :
10     CONTINUE
20 CONTINUE
```

As shown in the second example, each loop must have a separate label. Additional, but redundant, DO syntax is described in Section 11.3.2 (and Appendix C.2). The full DO construct syntax is given in Appendix B.

Finally, it should be noted that many short DO-loops can be expressed alternatively, and better, in the form of array expressions and assignments. However, this is not always possible, and a particular danger to watch for is where one iteration of the loop depends upon a previous one. Thus, the loop

```
    DO I = 2, N
       A(I) = A(I-1) + B(I)
    END DO
```

*cannot* be replaced by the statement

```
    A(2:N) = A(1:N-1) + B(2:N)      !Beware
```

## 4.6 Summary

In this chapter we have introduced the four main features by which the control in Fortran 90 code may be programmed — the GO TO statement, the IF statement and construct, the CASE construct and the DO construct. The effective use of these features is the key to sound code. Of these features, the CASE construct is new to Fortran, and the DO construct was formerly limited to the labelled form ending on a CONTINUE statement.

We have touched upon the concept of a *program unit* as being like the chapter of a book. Just as a book may have just one chapter, so a complete program may consist of just one program unit, which is known as a *main program*. In its simplest form it consists of a series of statements of the kinds we have been dealing with so far, and terminates with an END statement, which acts as a signal to the computer to stop processing the current program.

```
!   Print a conversion table of the Farenheit and Celsius
!   temperature scales between specified limits.
!
    REAL CELSIUS, FARENHEIT
    INTEGER LOW_TEMP, HIGH_TEMP, TEMPERATURE
    CHARACTER SCALE
!
READ_LOOP:    DO
!
!   Read SCALE and limits
      READ *, SCALE, LOW_TEMP, HIGH_TEMP
!
!   Check for valid data
      IF (SCALE /= 'C' .AND. SCALE /= 'F') EXIT READ_LOOP
!
!   Loop over the limits
      DO  TEMPERATURE = LOW_TEMP, HIGH_TEMP
!
!   Choose conversion formula
        SELECT CASE (SCALE)
        CASE ('C')
          CELSIUS = TEMPERATURE
          FARENHEIT = 9./5.*CELSIUS + 32.
        CASE ('F')
          FARENHEIT = TEMPERATURE
          CELSIUS = 5./9.*(FARENHEIT-32.)
        END SELECT
!
!   Print table
        PRINT *, CELSIUS, ' Degrees C correspond to',          &
                FARENHEIT, ' Degrees F'
      END DO
    END DO READ_LOOP
!
!   Termination
    PRINT *, ' End of valid data'
    END
C  90    100
F  20    32
*  0    0
```

Figure 12.

In order to test whether a program unit of this type works correctly, we need to be able to output, to a terminal or printer, the values of the computed quantities. This topic will be fully explained in Chapter 9, and for the moment we need to know only that this can be achieved by a statement of the form

```
PRINT * , ' VAR1 = ', VAR1 , ' VAR2 = ', VAR2
```

which will output a line such as

```
VAR1 = 1.0   VAR2 = 2.0
```

Similarly, input data can be read by statements like

```
READ *, VAL1, VAL2
```

This is sufficient to allow us to write simple programs like that in Figure 12, which outputs the converted values of a temperature scale between specified limits.  Valid inputs are shown at the end of the example.

## Exercises

**1.** Write a program which
   a) defines an array to have 100 elements;
   b) assigns to the elements the values 1,2,3,....100;
   c) reads two integer values in the range 1 to 100;
   d) reverses the order of the elements of the array in the range specified by the two values.

**2.** The first two terms of the Fibonacci series are both 1, and all subsequent terms are defined as the sum of the preceding two terms.  Write a program which reads an integer value LIMIT and which computes and prints the first LIMIT terms of the series.

**3.** The coefficients of successive orders of the binomial expansion are shown in the normal Pascal triangle form as

```
            1
          1   1
        1   2   1
      1   3   3   1
    1   4   6   4   1
            etc.
```

Write a program which reads an integer value LIMIT and prints the first LIMIT lines of this Pascal triangle.

**4.** Define a character variable of length 80.  Write a program which reads a value for this variable.  Assuming that each character in the variable is alphabetic, write code which sorts them into alphabetic order, and prints out the frequency of occurrence of each letter.

**5.** Write a program to read an integer value LIMIT and print the first LIMIT prime numbers, by any method.

**6.** Write a program which reads a value X, and calculates and prints the corresponding value X/(1.+X). The case X=−1. should produce an error message and be followed by an attempt to read a new value of X.

**7.** Given a chain of entries of the type ENTRY of Section 2.13, modify the code in Figure 9 (Section 4.5) so that it removes the entry with index 10, and makes the previous entry point to the following entry.

# 5. PROGRAM UNITS AND PROCEDURES

## 5.1 Introduction

As we saw in the previous chapter, it is possible to write a complete Fortran program as a single unit, and indeed this was a common practice in the early days of programming. Nowadays, it is generally considered to be preferable to break the program down into manageable units. Each such *program unit* corresponds to a program task that can be readily understood and, ideally, can be written and tested in isolation. We will discuss the three kinds of program unit, the main program, external subprogram, and module.

A complete program must, as a minimum, include one *main program*. This may contain statements of the kinds that we have met so far in examples, but normally its most important statements are invocations or *calls* to subsidiary programs known as *subprograms*. A subprogram defines a *function* or a *subroutine*. They differ in that a function returns a single value and usually does not alter the values of its arguments (so that it represents a function in the mathematical sense), whereas a subroutine usually performs a more complicated task, returning several results through its arguments and by other means. Functions and subroutines are known collectively as *procedures*.

There are various kinds of subprograms. A subprogram may be a program unit in its own right, in which case it is called an *external subprogram* and defines an *external procedure*. External procedures may also be defined by means other than Fortran (usually assembly language). A subprogram may be a member of a collection in a program unit called a *module*, in which case we call it a *module subprogram* and it defines a *module procedure*. A subprogram may be placed inside a module subprogram, an external subprogram, or a main program, in which case we call it an *internal subprogram* and it defines an *internal procedure*. Internal subprograms may not be nested, that is they may not contain further subprograms, and we expect them normally to be short sequences of code, say up to about twenty lines. We illustrate the nesting of subprograms in program units in Figure 13. If a program unit or subprogram contains a subprogram, it is called the *host* of that subprogram.

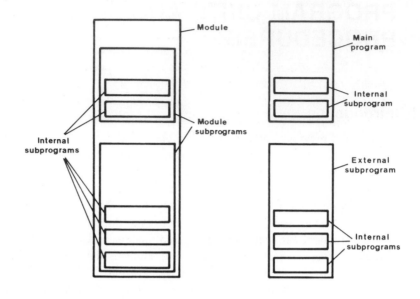

Figure 13. Nesting of subprograms in program units.

Besides containing a collection of subprograms, a module may contain data definitions, derived type definitions, interface blocks (Section 5.11), and namelist groups (Section 9.10). This collection may be expected to provide facilities associated with some particular task, such as providing matrix arithmetic, a library facility, or a data base. It may sometimes be large.

In this chapter we will describe program units and the statements that are associated with them. Within a complete program, they may appear in any order.

## 5.2  Main program

Every complete program must have one, and only one, main program. Optionally, it may contain calls to subprograms. A main program has the following form:

[PROGRAM *program-name*]
    [*specification-stmts*]
    [*executable-stmts*]
[CONTAINS
    *internal-subprograms*]
END [PROGRAM [*program-name*]]

The PROGRAM statement is optional, but we recommend its use. The *program-name* may be any valid Fortran name such as MODEL. The only non-optional statement is the END statement which has two purposes. It acts as a signal to the compiler that it has reached the end of the program unit and, when executed, it causes the complete program to stop. If it includes *program-name*, this must be the name on the PROGRAM statement.

The specification statements define the environment for the executable statements. So far, we have met the type declaration statement (INTEGER, REAL, COMPLEX, LOGICAL, CHARACTER, and TYPE(*type-name*)) that specifies the type and other properties of the entities that it lists, and the type definition block (bounded by TYPE *type-name* and END TYPE statements). We will meet other specification statements in this and the next two chapters.

The executable statements specify the actions that are to be performed. So far, we have met the assignment statement, the pointer assignment statement, the GO TO statement, the IF statement and construct, the DO and CASE constructs, and the READ and PRINT statements. We will meet other executable statements in this and later chapters. Execution of a program always commences with the first executable statement of the main program.

The CONTAINS statement flags the presence of one or more internal subprograms. We will describe internal subprograms in Section 5.6. If the execution of the last statement ahead of the CONTAINS statement does not result in a branch, control passes over the internal subprograms to the END statement and the program stops. The END statement may be labelled and may be the target of a branch from one of the executable statements. If such a branch is taken, again the program stops.

## 5.3  The STOP statement

Another way to stop program execution is to execute a STOP statement. This statement may be labelled, may be part of an IF statement, and is an executable statement that may appear in the main program or any subprogram. A well-designed program normally returns control to the main program for program termination, so the STOP statement should appear there. However, in applications where several STOP statements appear in various places in a complete program, it is possible to distinguish which of the STOP statements has caused the termination by adding to each one an *access code* consisting of a default character constant or a string of up to five digits whose leading zeros are not significant. This might be used by a given processor to indicate the origin of the STOP in a message. Examples are

```
STOP
STOP 'Incomplete data. Program terminated.'
STOP 12345
```

## 5.4 External subprograms

External subprograms are called from a main program or elsewhere, usually to perform a well-defined task within the framework of a complete program. Apart from the leading statement, they have a form that is very like that of a main program:

> *subroutine-stmt*
>   [*specification-stmts*]
>   [*executable-stmts*]
> [CONTAINS
>   *internal-subprograms*]
> END [SUBROUTINE [*subroutine-name*]]

or

> *function-stmt*
>   [*specification-stmts*]
>   [*executable-stmts*]
> [CONTAINS
>   *internal-subprograms*]
> END [FUNCTION [*function-name*]]

The CONTAINS statement plays exactly the same role as within a main program (see Section 5.2). The effect of executing an END statement in a subprogram is to return control to the caller, rather than to stop execution.

The simplest form of external subprogram defines a subroutine without any arguments and has a *subroutine-stmt* of the form

    SUBROUTINE subroutine-name

Such a subprogram is useful when a program consists of a sequence of distinct phases, in which case the main program consists of a sequence of CALL statements that invoke the subroutines as in the example

```
PROGRAM GAME          ! Main program to control a card game
    CALL SHUFFLE      ! First shuffle the cards.
    CALL DEAL         ! Now deal them.
    CALL PLAY         ! Play the game.
    CALL DISPLAY      ! Display the result.
END PROGRAM GAME      ! Cease execution.
```

However, there is something missing from this program — there is no sign of any flow of information between the subroutines. How does PLAY know which cards DEAL has dealt?  There are, in fact, two methods by which information may be passed.  The first is via data held in a module (Section 5.5) and accessed by the subprograms, and the second is via arguments (Section 5.7) in the procedure calls.

## 5.5  Modules

The third type of program unit, the module, is useful for global data and is important for the packaging of derived data types.  It has the form

MODULE *module-name*
  [ *specification-stmts*]
[CONTAINS
  *module-subprograms*]
END [MODULE [*module-name*]]

In its simplest form, the body consists only of data specifications.  For example

```
MODULE STATE
   INTEGER, DIMENSION(52) :: CARDS
END MODULE STATE
```

might hold the state of play of the game of Section 5.4.  It is accessed by the statement

```
USE STATE
```

appearing at the beginnings of the main program GAME and subprograms SHUFFLE, DEAL, PLAY, and DISPLAY.  The array CARDS is set by SHUFFLE to contain the integer values 1 to 52 in a random order, where each integer value corresponds to a pre-defined playing card.  For instance, 1 might stand for the ace of clubs, 2 for the two of clubs, etc. up to 52 for the king of spades.  The array CARDS is changed by the subroutines DEAL and PLAY, and finally accessed by subroutine DISPLAY.

```
MODULE INTERVAL_ARITHMETIC
    TYPE INTERVAL
        REAL LOWER, UPPER
    END TYPE INTERVAL
    INTERFACE OPERATOR(+)
        MODULE PROCEDURE ADD_INTERVALS
    END INTERFACE
    :
CONTAINS
    FUNCTION ADD_INTERVALS(A,B)
        TYPE(INTERVAL) ADD_INTERVALS, A, B
        ADD_INTERVALS%LOWER = A%LOWER + B%LOWER
        ADD_INTERVALS%UPPER = A%UPPER + B%UPPER
    END FUNCTION ADD_INTERVALS
    :
END MODULE INTERVAL_ARITHMETIC
```

Figure 14.

A very useful role for modules is to contain definitions of types and their associated operators. For example, a module might contain the type INTERVAL of Section 3.8, as shown in Figure 14. Given this module, any program unit needing this type and its operators need only include the statement

```
USE INTERVAL_ARITHMETIC
```

at the head of its specification statements.

A module subprogram has exactly the same form as an external subprogram, except that FUNCTION or SUBROUTINE *must* be present on the END statement, so there is no need for a separate description. It always has access to other entities of the module, including the ability to call other subprograms of the module, rather as if it contained a USE statement for its module.

A module may contain USE statements that access other modules. It must not access itself directly or indirectly through a chain of USE statements, for example A accessing B and B accessing A. No ordering of modules is required by the standard, but we expect normal practice to be to set up a module (say as part of a library) and then use it, so that it would be natural to access only preceding modules. We recommend this practice, which will make it impossible for a module to access itself through other modules.

It is possible within a module to specify that some of the entities are private to it and cannot be accessed from other program units. Also there are forms of the USE statement that allow access to only part of a module and forms that allow renaming of the entities accessed. These features will be explained in Sections 7.6 and 7.10. For the present, we assume that the whole module is accessed without any renaming of the entities in it.

Besides data definitions, type definitions, subprograms, and inteface blocks, a module may contain namelist groups (Section 7.15). The ability to make single definitions of interface blocks will be seen to be important in the context of constructing large libraries of reusable software.

## 5.6 Internal subprograms

We have seen that internal subprograms may be defined inside main programs and external subprograms, and within module subprograms. They have the form

*subroutine-stmt*
    [*specification-stmts*]
    [*executable-stmts*]
END SUBROUTINE [*subroutine-name*]

or

*function-stmt*
    [*specification-stmts*]
    [*executable-stmts*]
END FUNCTION [*function-name*]

that is, the same form as a module subprogram, except that they may not contain further internal subprograms. It automatically has access to all the host's entities, including the ability to call its other internal subprograms. A single internal subprogram, or a group of them, must be preceded by a CONTAINS statement in the host.

In the rest of this chapter, we describe several properties of subprograms that apply to external, module, and internal subprograms. We therefore do not need to describe internal subprograms separately.

## 5.7 Arguments of procedures

Procedure arguments provide an alternative means for two program units to access the same data. Returning to our card game example, instead of placing the array CARDS in a module, we might declare it in the main program and pass it as an actual argument to each subprogram, as shown in Figure 15.

```
PROGRAM GAME                 ! Main program to control a card game
   INTEGER, DIMENSION(52) :: CARDS
   CALL SHUFFLE(CARDS)       ! First shuffle the cards.
   CALL DEAL(CARDS)          ! Now deal them.
   CALL PLAY(CARDS)          ! Play the game.
   CALL DISPLAY(CARDS)       ! Display the result.
END PROGRAM GAME             ! Cease execution.
```

Figure 15.

Each subroutine receives CARDS as a dummy argument. For instance
SHUFFLE has the form shown in Figure 16.

```
SUBROUTINE SHUFFLE(CARDS)
   ! Subroutine that places the values 1 to 52 in CARDS
   ! in random order.
   INTEGER, DIMENSION(52) :: CARDS
   ! Statements that fill CARDS
   :
END SUBROUTINE SHUFFLE   ! Return to caller.
```

Figure 16.

We can, of course, imagine a card game in which DEAL is going to deal
only three cards to each of four players. In this case, it would be a waste of
time for SHUFFLE to prepare a deck of 52 cards when only the first 12 cards
are needed. This can be achieved by requesting SHUFFLE to limit itself to a
number of cards that is transmitted in the calling sequence thus:

```
CALL SHUFFLE(3*4, CARDS)
```

Inside SHUFFLE, we would define the array to be of the given length and the
algorithm to fill CARDS would be contained in a DO construct with this
number of iterations, as shown in Figure 17.

```
SUBROUTINE SHUFFLE(NCARDS, CARDS)
   INTEGER NCARDS
   INTEGER, DIMENSION(NCARDS) :: CARDS
   DO ICARD = 1, NCARDS
      :
      CARDS(ICARD) = ...
   END DO
END SUBROUTINE SHUFFLE
```

Figure 17.

We have seen how it is possible to pass an array and a constant expression between two program units. An actual argument may be any variable or expression (or a procedure name, see Section 5.12). Each dummy argument of the called procedure must agree with the corresponding actual argument in type, type parameters, and shape (the requirements on character length and shape agreement are relaxed in Chapter 11). However, the names do not have to be the same. For instance, if two decks had been needed, we might have written the code thus:

```
PROGRAM GAME
    INTEGER, DIMENSION(52) :: ACARDS, BCARDS
    CALL SHUFFLE(ACARDS)        ! First shuffle the A deck.
    CALL SHUFFLE(BCARDS)        ! Next shuffle the B deck.
    :
END PROGRAM GAME
```

The important point is that subprograms can be written independently of one another, the association of the dummy arguments with the actual arguments occurring each time the call is executed. We can imagine SHUFFLE being used in other programs which use other names. In this manner, libraries of subprograms may be built up.

Being able to have different names for actual and dummy arguments provides a useful flexibility, but it should only be used when it is actually needed. When the same name can be used, the code is more readable.

## 5.7.1  Pointer arguments

A dummy argument is permitted to have the attribute POINTER. In this case, the actual argument must also have the attribute POINTER. When the subprogram is invoked, the rank of the actual argument must match that of the dummy array, and its pointer association status is passed to the dummy array. On return, the actual argument takes its pointer association status from that of the dummy array, but note that the target may become undefined when the return is executed, for example if it is a dummy argument with the TARGET attribute (Section 5.7.3). The INTENT attribute (Section 5.9) would be ambiguous in this context, since it might refer to the pointer association status alone or to both the pointer association status and the value of its target; it is not allowed to be specified.

In the case of a module or internal procedure, the compiler knows when the dummy argument is a pointer. In the case of an external or dummy procedure, the compiler assumes that the dummy argument is not a pointer unless it is told otherwise in an interface block (Section 5.11).

A pointer actual argument is also permitted to correspond to a nonpointer dummy argument. In this case, the pointer must have a target and the target is associated with the dummy argument.

## 5.7.2 Restrictions on actual arguments

There are two important restrictions on actual arguments, which are designed to allow the compiler to optimize on the assumption that the dummy arguments are distinct from each other and from other entities that are accessible within the procedure. For example, a compiler may arrange for an array to be copied to a local variable on entry, and copied back on return. While an actual argument is associated with a dummy argument:

i) no action may be taken that affects the value or availability of the actual argument, except through the dummy argument, and

ii) if any part of the actual argument is defined through the dummy argument, the actual argument may be referenced only through the dummy argument.

As an example of i), consider

```
CALL MODIFY(A(1:5), A(3:9))
```

Here, A(3:5) may not be changed through either dummy argument since this would violate the rule for the other argument. However, A(1:2) may be changed through the first argument and A(6:9) may be changed through the second. Another example is an actual argument that is an object being accessed from a module; here, the same object must not be accessed from the module by the procedure and redefined. A third example is a pointer that is nullified (Section 6.7) while still associated with the dummy argument.

As an example of ii), suppose an internal procedure call associates a host variable H with a dummy argument D. If D is defined during the call, then at no time during the call may H be referenced directly.

## 5.7.3 Arguments with the TARGET attribute

Because an implementation is permitted to copy actual arguments on entry to a procedure and copy them back on return, any pointers associated with an actual argument that has the TARGET attribute do not become associated with the corresponding dummy argument but remain associated with the actual argument. Similarly, if a dummy argument has the TARGET attribute, any pointer associated with it becomes undefined on return.

## 5.8 The RETURN statement

We saw in Section 5.2 that if the last executable statement in a main program is executed and does not cause a branch, the END statement is executed and the program stops. Similarly, if the last executable statement in a subprogram is executed and does not cause a branch, the END statement is executed and control returns to the point of invocation. Just as the STOP statement is an executable statement that provides an alternative means of stopping execution, so the RETURN statement provides an alternative means of returning control from a subprogram. It has the form

        RETURN

Like the STOP statement, this statement may be labelled, may be part of an IF statement, and is an executable statement. It must not appear among the executable statements of a main program.

## 5.9 Argument intent

In Figure 17, the dummy argument CARDS was used to pass information out from SHUFFLE and the dummy argument NCARDS was used to pass information in. A third possibility is for a dummy argument to be used for both. We can specify the intent on the type declaration statement for the argument, for example:

```
SUBROUTINE SHUFFLE(NCARDS, CARDS)
    INTEGER, INTENT(IN)                       :: NCARDS
    INTEGER, INTENT(OUT), DIMENSION(NCARDS) :: CARDS
```

For input-output arguments, INTENT(INOUT) may be specified.

   If a dummy argument is specified with INTENT(IN), it must not be redefined by the procedure, say by appearing on the left-hand side of an assignment or by being passed on as an actual argument to a procedure that redefines it. For the specification INTENT(INOUT), the corresponding actual argument must be a variable because the expectation is that it will be redefined by the procedure. For the specification INTENT(OUT), the corresponding actual argument must again be a variable; in this case, it becomes undefined on entry to the procedure because the intention is that it be used only to pass information out.

   If a function specifies a defined operator (Section 3.8), the dummy arguments must have intent IN. If a subroutine specifies defined assignment (Section 3.9), the first argument must have intent OUT or INOUT, and the second argument must have intent IN.

If a dummy argument has no intent, the actual argument may be a variable or an expression, but the actual argument must be a variable if the dummy argument is redefined. It has been traditional for Fortran compilers not to check this rule, since they usually compile each program unit separately. Breaching the rule can lead to program errors at execution time that are very difficult to find. We recommend that all dummy arguments be given a declared intent. Not only is this good documentation, but it allows compilers to make more checks at compile time.

If a dummy argument has the POINTER attribute, its INTENT is not allowed to be specified. This is because of the ambiguity of whether the intent applies to the target data object or to the pointer association.

## 5.10 Functions

Functions are similar to subroutines in many respects, but they are invoked within an expression and return a value that is used within the expression. For example, the subprogram in Figure 18 returns the distance between two points in space and the statement

```
IF (DISTANCE(A, C) > DISTANCE(B, C) ) THEN
```

invokes the function twice in the logical expression that it contains.

---

```
FUNCTION DISTANCE(P, Q)
   REAL DISTANCE
   REAL, INTENT(IN), DIMENSION(3) :: P, Q
   DISTANCE = SQRT((P(1)-Q(1))**2 + (P(2)-Q(2))**2+(P(3)-Q(3))**2)
   ! The intrinsic function SQRT is defined in Section 8.4.
END FUNCTION DISTANCE
```

---

Figure 18.

Note the type declaration for the function result. The result behaves just like a dummy argument with intent OUT. It is initially undefined, but once defined it may appear in an expression and it may be redefined. The type may also be defined on the FUNCTION statement thus:

```
REAL FUNCTION DISTANCE(P, Q)
```

It is permissible to write functions that change the values of their arguments, modify values in modules, or perform input-output operations. However, these are known as *side-effects* and conflict with good programming practice. Where they are needed, a subroutine should be used. It is reassuring to know that when a function is called, nothing else goes on 'behind the scenes', and it may be very helpful to an optimizing compiler, particularly for internal and module subprograms.

A function result may be a pointer, which is very useful when the size of the result depends on a calculation in the function itself. The result is initially undefined. Within the function, it must become associated or defined as disassociated. We expect the function reference usually to be such that a pointer assignment takes place for the result, that is, the reference occurs as the right-hand side of a pointer assignment (Section 3.12) or as a pointer component of a structure constructor. If the reference occurs as a primary of an expression, the result must be associated with a target that is defined, and the value of the target is used for the evaluation of the expression.

The value returned by a nonpointer function must always be defined.

## 5.10.1 Prohibited side-effects

In order to assist an optimizing compiler, the standard prohibits reliance on certain side-effects. It specifies that it is not necessary for a processor to evaluate all the operands of an expression if the value of the expression can be determined otherwise. For example, in evaluating

```
X>Y .OR. L(Z)
```

where X, Y, and Z are real and L is a logical function, the function reference need not be made if X is greater than Y. Since some processors will make the call and others will not, any variable (for example Z) that is redefined by the function becomes undefined following such expression evaluation. Similarly, it is not necessary for a processor to evaluate any subscript or substring expressions for an array of zero size or character object of zero character length.

Another prohibition is that a function reference must not redefine the value of a variable that appears in the same statement or affect the value of another function reference in the same statement. For example, in

```
D = MAX(DISTANCE(P,Q), DISTANCE(Q,R))
```

DISTANCE is required not to redefine its arguments. This rule allows any expressions that are arguments of a single procedure call to be evaluated in any order. With respect to this rule, an IF statement,

IF (*lexpr*) stmt

is treated as the equivalent IF construct

IF (*lexpr*) THEN
   *stmt*
END IF

and the same is true for the WHERE statement (Section 6.8).

## 5.11  Explicit and implicit interfaces

A call to an internal subprogram must be from a statement within the same program unit. It may be assumed that the compiler will process the program unit as a whole and will therefore know all about any internal subprogram. In particular, it will know about its *interface*, that is whether it defines a function or a subroutine, the names and properties of the arguments, and the properties of the result if it defines a function. This, for example, permits the compiler to check whether the actual and dummy arguments match in the way that they should. We say that the interface is *explicit*.

A call to a module subprogram must either be from another statement in the module or from a statement following a USE statement for the module. In both cases, the compiler will know all about the subprogram, and again we say that the interface is explicit. Similarly, intrinsic procedures (Chapter 8) always have explicit interfaces.

When compiling a call to an external or dummy procedure, the compiler normally does not have a mechanism to access its code. We say that the interface is *implicit*. To specify that a name is that of an external or dummy procedure, the EXTERNAL statement is available. It has the form

   EXTERNAL *external-name-list*

and appears with other specification statements, after any USE or IMPLICIT statements and before any executable statements. The type and type parameters of a function with an implicit interface are specified by a type declaration statement for the function name or by the rules of implicit typing (Section 7.2) applied to the name.

The EXTERNAL statement merely specifies that each *external-name* is the name of an external or dummy procedure. It does not specify the interface, which remains implicit. However, a mechanism is provided in Fortran 90 for the interface to be specified. It may be done through an interface block of the form

   INTERFACE
      *interface-body*
   END INTERFACE

Normally, the *interface-body* is an exact copy of the subprogram without its executable statements and internal subprograms. However,

- the names of the arguments may be changed,

- any specification that does not give information about a dummy argument or function result may be omitted (for example, a declaration of a local variable), and

- the same information may be given by a different combination of statements.[8]

An *interface-body* may be provided for a call to an external procedure defined by means other than Fortran (usually assembly language).

Naming a procedure in an EXTERNAL statement or giving it an interface body ensures that it is an external or dummy procedure and makes any intrinsic procedure having the same name unavailable. We strongly recommend the practice for external procedures, since otherwise the processor is permitted to interpret the name as that of an intrinsic procedure. It is needed for portability since processors are permitted to provide additional intrinsics.

The interface block is placed in a sequence of specification statements and this suffices to make the interface explicit. Perhaps the most convenient way to do this is to place the interface block among the specification statements of a module and then USE the module. We imagine subprogram libraries being written as sets of external subprograms which are precompiled and whose interfaces are collected into modules. This keeps the modules of modest size. Note that if a procedure is accessible in a scoping unit, its interface is either explicit or implicit there. An external procedure may have an explicit interface in some scoping units and an implicit interface in others.

Interface blocks may also be used to allow procedures to be called as defined operators (Section 3.8), called as defined assignments (Section 3.9), or called under a single generic name. We therefore defer description of the full generality of the interface block until Section 5.18, where overloading is discussed.

An explicit interface is required to invoke a procedure with a pointer or target dummy argument or a pointer function result, and is required for several useful features that we will meet later in this and the next chapter. It is needed so that the processor can make the appropriate linkage. Explicit interfaces are also desirable because of the additional security that they provide. It

---

[8] A practice that is permitted by the standard, but which we do not recommend, is for a dummy argument to be declared implicitly as a procedure by invoking it in an executable statement. If the subprogram has such a dummy procedure, the interface will need an EXTERNAL statement for that dummy procedure.

is straightforward to ensure that all interfaces are explicit and we recommend the practice.

## 5.12 Procedures as arguments

So far, we have taken the actual arguments of a procedure invocation to be variables and expressions, but another possibility is for them to be procedures. Let us consider the case of a library subprogram to perform function minimization. It needs to receive the user's function, just as the subroutine SHUFFLE in Figure 17 needs to receive the required number of cards. The library code might look like the code in Figure 19. Notice the way the procedure argument is declared by an interface block playing a similar role to that of the type declaration statement for a data object. Although such an interface block is not required, we recommend its use. As a minimum, the procedure name must be declared in an EXTERNAL statement.

```
   REAL FUNCTION MINIMUM(A, B, FUNC)
! Returns the minimum value of the function FUNC(X) in the
! interval (A,B).
     REAL, INTENT(IN) :: A, B
     INTERFACE
        REAL FUNCTION FUNC(X)
           REAL, INTENT(IN) :: X
        END FUNCTION FUNC
     END INTERFACE
     REAL F,X
     :
     F = FUNC(X)    ! Invocation of the user function.
     :
   END FUNCTION MINIMUM
```

Figure 19.

Just as the type and shape of actual and dummy data objects must agree, so must the properties of the actual and dummy procedures. The agreement is exactly as for the agreement between a procedure and an interface body for that procedure (see Section 5.11). It would make no sense to specify an INTENT attribute (Section 5.9) for a dummy procedure, and this is not permitted.

On the user side, the code may look like that in Figure 20. Notice that the structure is rather like a sandwich: user-written code invokes the library code which in turn invokes user-written code.

```
PROGRAM MAIN
   REAL A, B, F
   INTERFACE
      REAL FUNCTION FUN(X)
         REAL, INTENT(IN) :: X
      END FUNCTION FUN
   END INTERFACE
   F = MINIMUM(1.0, 2.0, FUN)
   :
END PROGRAM
REAL FUNCTION FUN(X)
   :
END FUNCTION FUN
```

Figure 20.

The procedure that is passed must be an external or module procedure and its specific name must be passed when it also has a generic name (Section 5.18). Internal procedures are not permitted because it is anticipated that they may be implemented quite differently (for example, by in-line code), and because of the need to identify the depth of recursion when the host is recursive (Section 5.16) and the procedure involves host variables.

## 5.13 Keyword and optional arguments

In practical applications, argument lists can get long and many of the arguments may often not be needed. For example, a subroutine for constrained minimization might have the form

```
SUBROUTINE MINCON(N, F, X, UPPER, LOWER,                    &
              EQUALITIES, INEQUALITIES, CONVEX, XSTART)
```

On many calls, there may be no upper bounds, or no lower bounds, or no equalities, or no inequalities, or it may not be known whether the function is convex, or a sensible starting point may not be known. All the corresponding dummy arguments may be declared OPTIONAL (see also Section 7.8). For instance the bounds might be declared by the statement

```
REAL, OPTIONAL, DIMENSION(N) :: UPPER,LOWER
```

If the first four arguments are the only wanted ones, we may use the statement

```
CALL MINCON(N, F, X, UPPER)
```

but usually the wanted arguments are scattered.  In this case, we may follow a (possibly empty) ordinary positional argument list for leading arguments by a keyword argument list, as in the statement

```
CALL MINCON(N, F, X, EQUALITIES=Q, XSTART=X0)
```

The keywords are the dummy argument names and there must be no further positional arguments after the first keyword argument.

This example also illustrates the merits of both positional and keyword arguments as far as readability is concerned.  A small number of leading positional arguments (for example, N, F, X) are easily linked in the reader's mind to the corresponding dummy arguments.  Beyond this, the keywords are very helpful to the reader in making these links.  We recommend their use for long argument lists even when there are no gaps caused by optional arguments that are not present.

A non-optional argument must appear exactly once, either in the positional list or in the keyword list.  An optional argument may appear at most once, either in the positional list or in the keyword list.  An argument must not appear in both lists.

The called subprogram needs some way to detect whether an argument is present so that it can take appropriate action when it is not.  This is provided by the intrinsic function PRESENT (see Section 8.2).  For example

```
PRESENT(XSTART)
```

returns the value .TRUE. if the current call has provided a starting point and .FALSE. otherwise.  When it is absent, the subprogram might use a random number generator to provide a starting point.

A slight complication occurs if an optional dummy argument is used within the subprogram as an actual argument in a procedure invocation.  For example, our minimization subroutine might start by calling a subroutine that handles the corresponding equality problem by the call

```
CALL MINEQ(N, F, X, EQUALITIES, CONVEX, XSTART)
```

In such a case, an absent optional argument is also regarded as absent in the second-level subprogram.  For instance, when CONVEX is absent in the call of MINCON, it is regarded as absent in MINEQ too.  Such absent arguments may be propagated through any number of levels of calls.

Since the compiler will not be able to make the appropriate associations unless it knows the keywords (dummy argument names), the interface must be explicit (Section 5.11) if any of the dummy arguments are optional or keyword arguments are in use.  Note that an interface block may be provided for an external procedure to make the interface explicit. In all cases where an interface block is provided, it is the names of the dummy arguments in the block that are used to resolve the associations.

## 5.14 Scope of labels

Execution of the main program or a subprogram always starts at its first executable statement and any branching always takes place from one of its executable statements to another. Indeed, each subprogram has its own independent set of labels. This includes the case of a host subprogram with several internal subprograms. The same label may be used in the host and the internal subprograms without ambiguity.

This is our first encounter with *scope*. The scope of a label is a main program or a subprogram, excluding any internal subprograms that it contains. The label may be used unambiguously anywhere among the executable statements of its scope. Notice that the host END statement may be labelled and be a branch target from a host statement, that is the internal subprograms leave a hole in the scope of the host.

## 5.15 Scope of names

In the case of a named entity, there is a similar set of statements within which the name may always be used to refer to the entity. Here, type definitions and interface blocks as well as subprograms can knock holes in scopes. This leads us to regard each program unit as consisting of a set of non-overlapping scoping units. A *scoping unit* is one of the following:

- a derived-type definition,

- a procedure interface body, excluding any derived-type definitions and interface bodies contained within it, or

- a program unit or subprogram, excluding derived-type definitions, interface bodies, and subprograms contained within it.

An example containing five scoping units is shown in Figure 21.

```
MODULE SCOPE1              ! Scope 1
    :                      ! Scope 1
CONTAINS                   ! Scope 1
    SUBROUTINE SCOPE2      ! Scope 2
        TYPE SCOPE3        ! Scope 3
            :              ! Scope 3
        END TYPE           ! Scope 3
        INTERFACE          ! Scope 3
            :              ! Scope 4
        END INTERFACE      ! Scope 3
        :                  ! Scope 2
    CONTAINS               ! Scope 2
        FUNCTION SCOPE5(...) ! Scope 5
            :              ! Scope 5
        END FUNCTION       ! Scope 5
    END SUBROUTINE         ! Scope 2
END MODULE                 ! Scope 1
```

Figure 21.

Once an entity has been declared in a scoping unit, its name may be used to refer to it in that scoping unit. An entity declared in another scoping unit is always a different entity even if it has the same name and exactly the same properties.[9] Each is known as a *local* entity. This is very helpful to the programmer, who does not have to be concerned about the possibility of accidental name clashes. Note that this is true for derived types, too. Even if two derived types have the same name and the same components, entities declared with them are treated as being of different types (see footnote).

A USE statement of the form

USE *module-name*

is regarded as a re-declaration of all the module entities inside the local scoping unit, with exactly the same names and properties. The module entities are said to be accessible by *use association*. Names of entities in the module may not be used to declare local entities (but see Section 7.10 for a description of further facilities provided by the USE statement when greater flexibility is required).

In the case of a derived-type definition, a module subprogram, or internal subprogram, the name of an entity in the host (including an entity accessed by use association) is similarly treated as being automatically re-declared with the

---

9  Apart from the effect of storage association, which is not discussed until Chapter 11 and whose use we strongly discourage.

same properties, provided no entity with this name is declared locally, is a local dummy argument or function result, or is accessed by use association. The host entity is said to be accessible by *host association*. For example, in the subroutine INNER of Figure 22, X is accessible by host association, but Y is a separate local variable and the Y of the host is inaccessible.

```
SUBROUTINE OUTER
   REAL X, Y
   :
CONTAINS
   SUBROUTINE INNER
      REAL Y
      Y = X + 1.
      :
   END SUBROUTINE INNER
END SUBROUTINE OUTER
```

Figure 22.

Note that the host does not have access to the local entities of any subroutine that it contains.

Host association does not extend to interface blocks. This allows an interface body to be constructed mechanically from the specification statements of an external procedure.

Within a scoping unit, each named data object, procedure, derived type, named construct, and namelist group (Section 7.15) must have a distinct name, with the one exception of generic names of procedures (to be described in Section 5.18). Note that this means that any appearence of the name of an intrinsic procedure in another rôle makes the intrinsic procedure inaccessible. Within a type definition, each component of the type, named constant, intrinsic procedure, and derived type must have a distinct name. Apart from these rules, names may be re-used. For instance, a name may be used for the components of two types, or the arguments of two procedures referenced with keyword calls.

The names of program units are *global*, that is available anywhere in a complete program. Each must be distinct from the others and from any of the local entities of the program unit.

At the other extreme, the DO variable of an implied-DO in a DATA statement (Section 7.5) or an array constructor (Section 6.13) has a scope that is just the implied-DO. It is different from any other entity with the same name.

## 5.16  Recursive procedures

Normally, a subprogram may not invoke itself, either directly or indirectly through a sequence of other invocations. However, if the leading statement is prefixed RECURSIVE, as in Figure 23, then it may be invoked recursively.

---

```
RECURSIVE FUNCTION INTEGRATE(F, BOUNDS)
    ! Integrate f(x) from BOUNDS(1) to BOUNDS(2)
    REAL INTEGRATE
    INTERFACE
        FUNCTION F(X)
            REAL F, X
        END FUNCTION F
    END INTERFACE
    REAL, DIMENSION(2), INTENT(IN) :: BOUNDS
    :
END FUNCTION INTEGRATE
```

---

Figure 23.

For example, suppose that it is desired to integrate a function $f$ of $x$ and $y$ over a rectangle. We might write a Fortran function in a module to get $x$ as an argument and $y$ from the module:

```
MODULE FUNC
    REAL YVAL
    REAL, DIMENSION(2) :: XBOUNDS, YBOUNDS
CONTAINS
    FUNCTION F(XVAL)
        REAL F, XVAL
        F = ...        ! Expression involving XVAL and YVAL
    END FUNCTION
END MODULE
```

then we can integrate over $x$ for a particular value of $y$ thus

```
FUNCTION FY(Y)
    USE FUNC
    REAL FY, Y
    YVAL = Y
    FY = INTEGRATE(F, XBOUNDS)
END
```

and we may then integrate over the whole rectangle thus

```
INTEGRATE(FY, YBOUNDS)
```

Note that INTEGRATE calls FY, which in turn calls INTEGRATE.

Each time a recursive procedure is invoked, a fresh set of local data objects is created, which ceases to exist on return. They consist of all data objects declared in its specification statements or declared implicitly (see Section 7.2), but excepting those with the DATA or SAVE attribute (see Sections 7.5 and 7.9). The interface is explicit within the procedure.

## 5.17  The RESULT clause

In the example of the previous section, the function did not call itself directly, so the function name could be used for the function result. When it does call itself, another name is needed. This is done by adding another clause to the FUNCTION statement as in the example

```
RECURSIVE FUNCTION INTEGRATE(F, BOUNDS) RESULT(INTEGRAL)
```

All specification is done in terms of the result name, that is the function name must not appear among the specification statements. In the executable statements, a reference to the function name is a recursive invocation of of the function and the result name must be used for the result variable.

The recursive function FACTORIAL in Figure 24 calculates $N!=N(N-1)...(1)$ and illustrates the use of a RESULT clause.

```
RECURSIVE FUNCTION FACTORIAL(N) RESULT(RES)
    INTEGER RES, N
    IF(N.EQ.1) THEN
        RES = 1
    ELSE
        RES = N*FACTORIAL(N-1)
    END IF
END
```

Figure 24.

If there is no RESULT clause, the function name is used for the result, and is not available for a recursive function call.

## 5.18 Overloading and generic interfaces

We saw in Section 5.11 how to use a simple interface block to provide an explicit interface to an external or dummy procedure. Another use is for overloading, that is being able to call several procedures by the same generic name. Here the interface block contains several interface bodies and the INTERFACE statement specifies the generic name. For example,

```
INTERFACE GAMMA
    FUNCTION SGAMMA(X)
        REAL (SELECTED_REAL_KIND( 6)) SGAMMA, X
    END
    FUNCTION DGAMMA(X)
        REAL (SELECTED_REAL_KIND(12)) DGAMMA, X
    END
END INTERFACE
```

permits both the functions SGAMMA and DGAMMA to be invoked using the generic name GAMMA.

A specific name for a procedure may be the same as its generic name. For example, the procedure SGAMMA could be renamed GAMMA without invalidating the interface block.

Furthermore, a generic name may be the same as another accessible generic name. In such a case, all the procedures that have this generic name may be invoked through it. This capability is important, since a module may need to extend the intrinsic functions such as SIN to a new type such as INTERVAL (Section 3.8).

If it is desired to overload a module procedure, the interface is already explicit so it is inappropriate to specify an interface body. Instead, the statement

MODULE PROCEDURE *procedure-name-list*

is included in the interface block in order to name the module procedures for overloading: if the functions SGAMMA and DGAMMA above were defined in a module, the statements

```
INTERFACE GAMMA
    MODULE PROCEDURE SGAMMA, DGAMMA
END INTERFACE
```

only would be required in any program unit which accesses the module and references the procedures by the generic name GAMMA.

Another form of overloading occurs when an inteface block specifies a defined operation (Section 3.8) or a defined assignment (Section 3.9) to *extend*

an intrinsic operation or assignment. The scope of the defined operation or
assignment is the scoping unit that contains the interface block, but it may be
accessed elsewhere by use or host association. If an intrinsic operator is
extended, the number of arguments must be consistent with the intrinsic form
(for example, it is not possible to define a unary *).

The general form of the interface block is thus

INTERFACE [ *generic-spec* ]
    [*interface-body* ]...
    [MODULE PROCEDURE *procedure-name-list*]...
END INTERFACE

where *generic-spec* is

*generic-name*,

OPERATOR(*defined-operator*), or

ASSIGNMENT(=)

A MODULE PROCEDURE statement is permitted only when a *generic-spec*
is present, and all the procedures must be accessible module procedures. An
interface body must not be provided for a procedure that already has an
explicit interface (that is, a module procedure, an internal procedure, a procedure defined by the host if it is a subprogram, or a procedure that already has
an interface body).

If OPERATOR is specified on the interface statement, all the procedures in
the block must be functions with one or two non-optional arguments having
the intent IN. If ASSIGNMENT is specified, all the procedures must be subroutines with two non-optional arguments, the first having intent OUT or
INOUT and the second intent IN. In order that invocations are always unambiguous, if two procedures have the same generic operator or both define
assignment, one must have a dummy argument that corresponds by position in
the argument list to a dummy argument of the other that has a different type,
different kind type parameter, or different rank.

All procedures that have a given generic name must be subroutines or all
must be functions. Again, any two must differ sufficiently for any invocation
to be unabiguous. The rule is that at least one of them must have a
non-optional dummy argument that both

- corresponds by position in the argument list to a dummy argument that is
  not present in the other, is present with a different type or kind type
  parameter, or is present with a different rank, and

- corresponds by name to a dummy argument that is not present in the other, is present with a different type or kind type parameter, or is present with a different rank.

Both rules are needed in order to cater for both keyword and positional dummy argument lists. For instance, the interface in Figure 25 would not otherwise be valid because the two functions are always distinguishable in a positional call, but not on a keyword call such as F(I=INT, X=POSN).

```
!  Example of a broken overloading rule
   INTERFACE F
      FUNCTION FXI(X,I)
         REAL FXI, X
         INTEGER I
      END
      FUNCTION FIX(I,X)
         REAL FIX, X
         INTEGER I
      END
   END INTERFACE
```

Figure 25.

If an intrinsic procedure is overloaded, these rules for ensuring unambiguous calls must include the intrinsic procedure when regarded as a collection of specific procedures, one for each allowed combination of type, kind type parameter, and rank for each argument.

## 5.19  Assumed character length

A character dummy argument may be declared with an asterisk for the value of the length type parameter, in which case it automatically takes the value from the actual argument. For example, a subroutine to sort the elements of a character array might be written thus

```
SUBROUTINE SORT(N,CHARS)
   INTEGER N
   CHARACTER(LEN=*), DIMENSION(N) :: CHARS
      :
END
```

If the length of the associated actual argument is needed within the procedure, the intrinsic function LEN (Section 8.6) may be invoked, as in Figure 26.

```
INTEGER FUNCTION COUNT (LETTER, STRING)
   CHARACTER (1), INTENT(IN) :: LETTER
   CHARACTER (*), INTENT(IN) :: STRING
!    Count the number of occurrences of LETTER in STRING
   COUNT = 0
   DO I = 1, LEN(STRING)
      IF (STRING(I:I) == LETTER) COUNT = COUNT + 1
   END DO
END FUNCTION COUNT
```

Figure 26.

An asterisk must not be used for a kind type parameter value. This is because a change of character length is analogous to a change of an array size and can easily be accommodated in the object code, whereas a change of kind probably requires a different machine instruction for every operation involving the dummy argument. A different version of the procedure would need to be generated for each possible kind value of each argument. The overloading feature (previous section) gives the programmer an equivalent functionality with explicit control over which versions are generated.

## 5.20  The SUBROUTINE and FUNCTION statements

We finish this chapter by giving the full syntax of the SUBROUTINE and FUNCTION statements, which have so far been explained through examples. The full syntax is

[RECURSIVE]                                              &
SUBROUTINE *subroutine-name* [([*dummy-argument-list*])]

and

[ *prefix*] FUNCTION *function-name* ([*dummy-argument-list*]) &
[RESULT( *result-name*) ]

where *prefix* is

*type* [RECURSIVE]

or

RECURSIVE [ *type*]

(for details of *type* see Section 7.13).

Each feature has been explained separately and the meanings are the same in the combinations allowed by the syntax.

## 5.21  Order of statements

We have now met examples of all the different classes of statement which the Fortran language contains, except for the FORMAT statement (in Chapter 9, in a class by itself).  Each time a new class of statement has been introduced, some mention has been made of the position in which it may appear, and these are summarized in Figure 27.  PARAMETER, DATA, and IMPLICIT statements are specification statements that are described in Chapter 7.

| PROGRAM, FUNCTION, SUBROUTINE, or MODULE Statement | | | |
|---|---|---|---|
| USE Statements | | | |
| FORMAT Statements | IMPLICIT NONE Statement | | |
| | PARAMETER Statements | IMPLICIT Statements | |
| | PARAMETER and DATA Statements | Derived-type Definitions, Interface Blocks, Type Declaration Statements, and Specification Statements | |
| | Executable Statements | | |
| CONTAINS Statement | | | |
| Internal Subprograms or Module Subprograms | | | |
| END Statement | | | |

Figure 27.  Order of statements.

## 5.22 Summary

A program consists of a sequence of program units, in any order. It must contain exactly one main program but may contain any number of modules and external subprograms. We have described each kind of program unit. Modules contain data definitions, type definitions, namelist groups, interface blocks, and module subprograms, all of which may be accessed in other program units with the USE statement.

Subprograms define procedures, which may be functions or subroutines. They may also be defined intrinsically (Chapter 8) and external procedures may be defined by means other than Fortran. We have explained how information is passed between program units and to procedures through argument lists and through the use of modules. Procedures may be called recursively provided they are suitably labelled.

The interface to a procedure may be explicit or implicit. If it is explicit, keyword calls may be made, and the procedure may have optional arguments. Interface blocks permit procedures to be invoked as operations or assignments, or by a generic name. The character lengths of dummy arguments may be assumed.

We have also explained about the scope of labels and Fortran names, and introduced the concept of a scoping unit.

Many of the features are new to Fortran: the internal subprogram, modules, the interface block, optional and keyword arguments, argument intent, pointer dummy arguments and function results, recursion, and overloading. These are powerful additions to the language, particularly in the construction of large programs and libraries.

## Exercises

**1.** A subroutine receives as arguments an array of values, $x$, and the number of elements in $x$, $n$. If the mean and variance of the values in $x$ are estimated by

$$mean = (\sum_{i=1}^{n} x(i))/n$$

and

$$variance = (\sum_{i=1}^{n} (x(i)-mean)^2)/(n-1)$$

write a subroutine which returns these calculated values as arguments. The subroutine should check for invalid values of $n$ ($\leq 1$).

**2.** A subroutine MATRIX_MULT multiplies together two matrices A and B, whose dimensions are I×J and J×K, respectively, returning the result in a matrix C, dimensioned I×K. Write MATRIX_MULT, given that each element of C is defined by

$$C(m,n) = \sum_{l=1}^{J} (A(m,l) \times B(l,n))$$

The matrices should appear as arguments to MATRIX_MULT.

**3.** The subroutine RANDOM_NUMBER (Section 8.15.2) returns a random number in the range 0.0 to 1.0, that is

CALL RANDOM_NUMBER(R)    ! 0≤R<1

Using this function, write the subroutine SHUFFLE of Figure 16

**4.** A character string consists of a sequence of letters. Write a function to return that letter of the string which occurs earliest in the alphabet, for example, the result of applying the function to 'DGUMVETLOIC' is 'C'.

**5.** Write an internal procedure to calculate the volume of a cylinder of radius $r$ and length $l$, $\pi r^2 l$, using as the value of $\pi$ the result of ACOS(−1.0), and reference it in a host procedure.

**6.** Choosing a simple card game of your own choice, and using the random number procedure (Section 8.15.2), write subroutines DEAL and PLAY of Section 5.4, using data in a module to communicate between them.

**7.** Objects of the intrinsic type CHARACTER are of a fixed length. Write a module containing a definition of a variable length character string type, of maximum length 80, and also the procedures necessary to:

i) assign a character variable to a string;

ii) assign a string to a character variable;

iii) return the length of a string;

iv) concatenate two strings.

# 6. ARRAY FEATURES

## 6.1 Introduction

In an era when an increasing number of computers have the hardware capability to perform operations on *vectors* of operands, it is self-evident that a numerically based language such as Fortran should have matching notational facilities. Such facilities provide not only a notational convenience for the programmer, but provide too an opportunity to extend the power of the language to manipulate arrays. In addition, a concise syntax makes the presence of array manipulation obvious to compilers which, especially on vector processors, are able to generate better optimized object code than was formerly possible with DO-loops. However, to achieve this gain new optimization techniques are required, for instance the ability to recognise that two or more consecutive array statements may, in some cases, be processed in a single loop at the object code level. These techniques are being progressively introduced.[10]

Arrays were introduced in Sections 2.10 to 2.13, their use in simple expressions and in assignments was explained in Sections 3.10 and 3.12, and they were used as procedure arguments in Chapter 5. These descriptions were deliberately restricted because Fortran 90 contains a very full set of array features whose complete description would have unbalanced those chapters. The purpose of this chapter is to describe the array features in detail, but without anticipating the descriptions of the array intrinsic procedures of Chapter 8; the rich set of intrinsic procedures should be regarded as an integral part of the array features.

## 6.2 Zero-sized arrays

It might be thought that an array would always have at least one element. However, such a requirement would force programs to contain extra code to deal with certain natural situations. For example, the code in Figure 28. solves a lower-triangular set of linear equations. When I has the value N, the sections have size zero, which is just what is required.

---

[10] A fuller discussion of this topic can be found in *Optimizing Supercompilers for Supercomputers*, M. Wolfe (Pitman, 1989).

```
DO I = 1,N
    X(I) = B(I) / A(I, I)
    B(I+1:N) = B(I+1:N) - A(I+1:N, I) * X(I)
END DO
```

Figure 28.

Fortran 90 allows arrays to have zero size in all contexts. Whenever a lower bound exceeds the corresponding upper bound, the array has size zero.

There are few special rules for zero-sized arrays because they follow the usual rules, though some care may be needed in their interpretation. For example, two zero-sized arrays of the same rank may have different shapes. One might have shape (0,2) and the other (0,3) or (2,0). Such arrays are not conformable and therefore may not be used as the operands of a binary operation. However, an array is always conformable with a scalar so the statement

   *zero-sized-array = scalar*

is valid and the scalar is 'broadcast to all the array elements', making this a 'do nothing' statement.

A zero-sized array is regarded as being defined always, because it has no values that can be undefined.

## 6.3 Assumed-shape arrays

Outside Chapter 11, we require that the shape of actual and dummy arguments agree, and so far we have achieved this by passing the extents of the array arguments as additional arguments. However, it is possible to require that the shape of the dummy array be taken automatically to be that of the actual array argument. Such an array is said to be an *assumed-shape* array. When the shape is declared by the DIMENSION clause, each dimension has the form

   [*lower-bound*]:

where *lower-bound* is an integer expression that may depend on module data or the other arguments (see Section 7.14 for the exact rules). If *lower-bound* is omitted, the default value is 1. Note that it is the shape that is passed, and not the upper and lower bounds. For example, if the actual array is A, declared thus:

```
REAL, DIMENSION(0:10, 0:20) :: A
```

and the dummy array is DA, declared thus:

```
REAL, DIMENSION(:, :) :: DA
```

then A(I,J) corresponds to DA(I+1,J+1); to get the natural correspondence, the lower bound must be declared:

```
REAL, DIMENSION(0:, 0:) :: DA
```

In order that the compiler knows that additional information is to be supplied, the interface must be explicit (Section 5.11) at the point of call. A dummy array with the POINTER attribute is not regarded as an an assumed-shape array because its shape is not necessarily assumed.

## 6.4 Automatic objects

A procedure with dummy arguments that are arrays whose size varies from call to call may also need local arrays whose size varies. A simple example is the array WORK in the subroutine to interchange two arrays that is shown in Figure 29.

```
SUBROUTINE SWAP(A, B)
   REAL, DIMENSION(:)        :: A, B
   REAL, DIMENSION(SIZE(A)) :: WORK
             ! SIZE provides the size of an array,
             ! and is defined in Section 8.12.
   WORK = A
   A = B
   B = WORK
END SUBROUTINE
```

Figure 29.

Arrays whose extents vary in this way are called *automatic arrays*, and are examples of *automatic data objects*. These are data objects whose declarations depend on the value of non-constant expressions, and are not dummy arguments. Implementations are likely to bring them into existence when the procedure is called and destroy them on return, maintaining them on a stack. The nonconstant expressions are limited to be specification expressions (Section 7.14).

The other way that automatic objects arise is through varying character length. The variable WORD2 in

```
SUBROUTINE EXAMPLE(WORD1)
   CHARACTER(LEN = *)          WORD1
   CHARACTER(LEN = LEN(WORD1)) WORD2
```

is an example. If a function result has varying character length, the interface must be explicit at the point of call because the compiler needs to know this. An array bound or the character length of an automatic object is fixed for the duration of each execution of the procedure and does not vary if the value of the specification expression varies or becomes undefined.

Some small restrictions on the use of automatic data objects appear in Sections 7.5, 7.9, and 7.15.

## 6.5  Elemental operations and assignments

We saw in Section 3.10 that an intrinsic operator can be applied to conformable operands, to produce an array result whose element values are the values of the operation applied to the corresponding elements of the operands. Such an operation is called *elemental*.

It is not essential to use operator notation to obtain this effect. Many of the intrinsic procedures (Chapter 8) are elemental and have scalar dummy arguments that may be called with array actual arguments provided all the array arguments have the same shape. For a function, the shape of the result is the shape of the array arguments. For example, we may find the square roots of all the elements of a real array thus:

```
A = SQRT(A)
```

For a subroutine, if any argument is array-valued, all the arguments with intent OUT or INOUT must be arrays.

Similarly, an intrinsic assignment may be used to assign a scalar to all the elements of an array, or to assign each element of an array to the corresponding element of an array of the same shape (Section 3.11). Such an assignment is also called *elemental*.

If a similar effect is desired for a defined operator, a function must be provided for each rank or pair of ranks for which it is needed. For example, the module in Figure 30 provides summation for scalars and rank-one arrays of intervals (Section 3.8). We leave it as an exercise for the reader to add definitions for mixing scalars and rank-one arrays.

Similarly, elemental versions of defined assignments must be provided explicitly.

```
MODULE INTERVAL_ADDITION
    TYPE INTERVAL
        REAL LOWER, UPPER
    END TYPE INTERVAL
    INTERFACE OPERATOR(+)
        MODULE PROCEDURE ADD00, ADD11
    END INTERFACE
CONTAINS
    FUNCTION ADD00 (A, B)
        TYPE (INTERVAL) ADD00, A, B
        ADD00%LOWER = A%LOWER + B%LOWER
        ADD00%UPPER = A%UPPER + B%UPPER
    END FUNCTION ADD00
    FUNCTION ADD11 (A, B)
        TYPE (INTERVAL), DIMENSION(:) :: A
        TYPE (INTERVAL), DIMENSION(SIZE(A)) :: B, ADD11
        ADD11%LOWER = A%LOWER + B%LOWER
        ADD11%UPPER = A%UPPER + B%UPPER
    END FUNCTION ADD11
END MODULE
```

Figure 30.

## 6.6  Array-valued functions

For simplicity in Section 5.10 we did not mention that a function may have an array-valued result, but we nevertheless used this language feature in Figure 30 since the interpretation is obvious.

In order that the compiler should know the shape of the result, the interface must be explicit (Section 5.11) whenever such a function is referenced. The shape is specified within the function definition by the DIMENSION attribute for the function name. Unless the function is a pointer, the bounds must be explicit expressions and they are evaluated on entry to the function.

An array-valued function is not necessarily elemental. For example, at the end of Section 3.10 we considered the type

```
TYPE MATRIX
    REAL ELEMENT
END TYPE MATRIX
```

Its scalar and rank-one operations might be as for reals, but for multiplying a rank-two array by a rank-one array, we might use the module function shown in Figure 31 to provide matrix by vector multiplication.

```
FUNCTION MULT(A, B)
   TYPE(MATRIX), DIMENSION(:, :)        :: A
   TYPE(MATRIX), DIMENSION(SIZE(A, 2)) :: B
                   ! SIZE is defined in Section 8.12
   TYPE(MATRIX), DIMENSION(SIZE(A, 1)) :: MULT
   INTEGER J, N
   MULT = 0.       ! A defined assignment from a real scalar to a
                   ! rank-one matrix.
   N = SIZE(A, 1)
   DO J = 1, SIZE(A, 2)
      MULT = MULT + A(1:N, J) * B(J) ! Uses defined operations for
                                     ! addition of two rank-one matrices
                                     ! and multiplication of a rank-one
                                     ! matrix by a scalar matrix.
   END DO
END FUNCTION
```

Figure 31.

## 6.7  Heap storage

There is an underlying assumption in Fortran 90 that the processor supplies a mechanism for managing heap storage.  The statements described in this section are the user interface to that mechanism.

### 6.7.1  Allocatable arrays

Sometimes an array is required to be of a size that is known only after some data have been read or some calculations performed.  An array with the POINTER attribute might be used for this purpose, but this is really not appropriate if the other properties of pointers are not needed.  Instead, an array that is not a dummy argument or function result may be given the ALLOCATABLE attribute by a statement such as

```
REAL, DIMENSION(:, :), ALLOCATABLE :: A
```

Such an array is called *allocatable*.  Its rank is specified when it is declared, but the bounds are undefined until an ALLOCATE statement such as

```
ALLOCATE(A(N, 0:N+1))     ! N of type integer
```

has been executed for it.  Its initial allocation status is 'not currently allocated' and it becomes allocated following successful execution of an ALLOCATE statement.

An important example is shown in Figure 32.  The array WORK is placed in a module and is allocated at the beginning of the main program to a size that depends on input data.  The array is then available throughout program execution in any subprogram that has a USE statement for WORK_ARRAY.

```
MODULE WORK_ARRAY
    INTEGER N
    REAL, DIMENSION(:,:,:), ALLOCATABLE :: WORK
END MODULE
PROGRAM MAIN
    USE WORK_ARRAY
    READ (INPUT, *) N
    ALLOCATE(WORK(N, 2*N, 3*N))
    :
```

Figure 32.

When an allocatable array A is no longer needed, it may be deallocated by execution of the statement

```
DEALLOCATE( A )
```

following which the array is 'not currently allocated'.  The DEALLOCATE statement is described in more detail in Section 6.7.3.

If it is required to make any change to the bounds of an allocatable array, the array must be deallocated and then allocated afresh.  Allocating an allocatable array that is already allocated, or deallocating an allocatable array that is not currently allocated, is an error.

An allocatable array that does not have the SAVE attribute (Section 7.9) may have a third allocation state: undefined.  Since an undefined allocatable array may not be referenced in any way, we recommend avoiding this state.  It occurs on return from a subprogram if the array is local to the subprogram or local to a module that is currently accessed only by the subprogram, and the array is allocated.  To avoid this situation, such an allocatable array must be explicitly deallocated before such a return.

## 6.7.2  ALLOCATE statement

We mentioned in Section 2.13 that the ALLOCATE statement can be used to give fresh storage for a pointer target directly.  The general form of the ALLOCATE statement is

ALLOCATE( *allocation-list* [, STAT=*stat*] )

where *allocation-list* is a list of allocations of the form

   *allocate-object* [( *array-bounds-list* )]

each *array-bound* has the form

   [ *lower-bound*:] *upper-bound*

and *stat* is a scalar integer variable that must not be part of an object being allocated.

If the STAT= specifier is present, *stat* is given either the value zero after a successful allocation or a positive value after an unsuccessful allocation (for example, if insufficient storage is available). If STAT= is absent and the allocation is unsuccessful, program execution stops.

Each *allocate-object* is an allocatable array or a pointer. It may have zero character length and in the case of a pointer may be a structure component.

Each *lower-bound* and each *upper-bound* is a scalar integer expression. The default value for the lower bound is 1. The number of *array-bound*s in a list must equal the rank of the *allocate-object*. They determine the array bounds, which do not alter if the value of a variable in one of the expressions changes subsequently. An array may be allocated to be of size zero, and it may have zero character length.

The bounds of all the arrays being allocated are regarded as undefined during the execution of the ALLOCATE statement, so none of the expressions that specify the bounds may depend on any of the bounds. For example,

```
ALLOCATE ( A(SIZE(B)), B(SIZE(A)) )   ! Illegal
```

or even

```
ALLOCATE ( A(N), B(SIZE(A)) )          ! Illegal
```

is not permitted, but

```
ALLOCATE ( A(N) )
ALLOCATE ( B(SIZE(A)) )
```

is valid. This restriction allows the processor to perform the allocations in a single ALLOCATE statement in any order.

In contrast to the case with an allocatable array, a pointer may be allocated a new target even if it is currently associated with a target. In this case, the previous association is broken. If the previous target was created by allocation, it becomes inaccessible unless another pointer is associated with it.

We expect linked-lists normally to be created by using a single pointer in an ALLOCATE statement for each node of the list, using pointer components of the allocated object at the node to hold the links from the node. We illustrate this by the addition of an extra nonzero element to the sparse vector held as a chain of entries of the type

```
TYPE ENTRY
   REAL VALUE
   INTEGER INDEX
   TYPE(ENTRY), POINTER :: NEXT
END TYPE ENTRY
```

of Section 2.3. The code in Figure 33 adds the new entry at the front of the chain. Note the importance of the last statement being a pointer assignment: the assignment

```
FIRST = CURRENT
```

would overwrite the old leading entry by the new one.

---

```
TYPE(ENTRY), POINTER :: FIRST, CURRENT
REAL FILL
INTEGER FILL_INDEX
:
ALLOCATE (CURRENT)
CURRENT = ENTRY (FILL, FILL_INDEX, FIRST)
FIRST => CURRENT
```

---

Figure 33.

## 6.7.3  DEALLOCATE statement

When an allocatable array or pointer target is no longer needed, its storage may be recovered by using the DEALLOCATE statement. Its general form is

DEALLOCATE ( *allocate-object-list* [, STAT=*stat*] )

where each *allocate-object* is an allocatable array that is allocated or a pointer that is associated with the whole of a target that was allocated through a pointer in an ALLOCATE statement. Here *stat* is a scalar integer variable that must not be deallocated by the statement. If STAT= is present, *stat* is given either the value zero after a successful execution or a positive value after an unsuccessful execution (for example, if a pointer is disassociated). If STAT= is absent and the deallocation is unsuccessful, program execution stops.

A danger in using the DEALLOCATE statement is that storage may be deallocated while pointers are still associated with the targets it held. Such pointers are left 'dangling' in an undefined state, and must not be reused until they are again associated with an actual target.

In order to avoid an accumulation of unused and unusable storage, all explicitly allocated storage should be explicitly deallocated when it is no longer required. This explicit management is required in order to avoid a potentially significant overhead on the part of the processor in handling arbitrarily complex allocation and reference patterns. Note also that the standard does not specify whether the processor recovers storage allocated through a pointer but no longer accessible through this or any other pointer.

### 6.7.4 NULLIFY statement

A pointer may be explicitly disassociated from its target by executing a NULLIFY statement.

The general form of the NULLIFY statement is

NULLIFY( *pointer-object-list* )

The statement is also useful for giving the disassociated status to an undefined pointer. An advantage of nullifying pointers rather than leaving them undefined is that they may then be tested by the intrinsic function ASSOCIATED (Section 8.2). For example, the end of the chain of Figure 33 might be flagged by a disassociated pointer.

### 6.8 WHERE statement and construct

It is often desired to perform an array operation only for certain elements, say those whose values are positive. The WHERE statement provides this facility. A simple example is

```
WHERE ( A > 0.0 ) A = 1.0/A    ! A is a real array
```

which reciprocates the positive elements of A and leaves the rest unaltered. The general form is

WHERE (*logical-array-expr*) *array-variable* = *array-expr*

The logical array expression *logical-array-expr* must have the same shape as *array-variable*. It is evaluated first and then just those elements of *array-expr* that correspond to elements of *logical-array-expr* that have the value true are evaluated and are assigned to the corresponding elements of *array-variable*.

All other elements of *array-variable* are left unaltered.  The assignment must not be a defined assignment.

A single logical array expression may be used for a sequence of array assignments all of the same shape.  The general form of this construct is

WHERE (*logical-array-expr*)
  *array-assignments*
END WHERE

The logical array expression *logical-array-expr* is first evaluated and then each array assignment is performed in turn, under the control of this mask.  If any of these assignments affect entities in *logical-array-expr*, it is always the value obtained when the WHERE statement is executed that is used as the mask.

Finally, the WHERE construct may take the form

WHERE (*logical-array-expr*)
  *array-assignments*
ELSEWHERE
  *array-assignments*
END WHERE

Here, the assignments in the first block of assignments are performed in turn under the control of *logical-array-expr* and then the assignments in the second block are performed in turn under the control of .NOT.*logical-array-expr*. Again, if any of these assignments affect entities in *logical-array-expr*, it is always the value obtained when the WHERE statement is executed that is used as the mask.  No array assignment in a WHERE construct may be a branch target statement.

A simple example of a WHERE construct is

```
WHERE (PRESSURE <= 1.0)
    PRESSURE = PRESSURE + INC_PRESSURE
    TEMP = TEMP + 5.0
ELSEWHERE
    RAINING = .TRUE.
END WHERE
```

where PRESSURE, INC_PRESSURE, TEMP, and RAINING are arrays of the same shape.

If a WHERE statement or construct masks an elemental function reference, the function is called only for the wanted elements.  For example,

```
WHERE ( A > 0 ) A = LOG(A)     ! LOG is defined in Section 8.4
```

would not lead to erroneous calls of LOG for negative arguments.

Apart from the elemental case, functions with array arguments are not permitted because masking would be inappropriate.

## 6.9  Array elements

In Section 2.10, we restricted the description of array elements to simple cases. In general, an array element is a scalar of the form

*part-ref* [*%part-ref*]...

where *part-ref* is

*part-name*[(*subscript-list*)]

At least one *part-ref* must have a *subscript-list*. The number of subscripts in each list must be equal to the rank of the array or array component, and each subscript must be a scalar integer expression whose value is within the bounds of its dimension of the array or array component. To illustrate this, take the type

```
TYPE TRIPLET
   REAL U
   REAL, DIMENSION(3)   :: DU
   REAL, DIMENSION(3,3) :: D2U
END TYPE TRIPLET
```

which was considered in Section 2.10. An array may be declared of this type:

```
TYPE(TRIPLET), DIMENSION(10,20,30) :: TAR
```

and

```
TAR(N,2,N*N)        ! N of type integer
```

is an array element. It is a scalar of type TRIPLET and

```
TAR(N, 2, N*N) % DU
```

is a real array with

```
TAR(N, 2, N*N) % DU(2)
```

as one of its elements.  It is also an element of the array TAR%DU(2).

If an array element is of type character, it may be followed by a substring reference:

(*substring-range*)

for example,

```
PAGE (K*K) (I+1:J-5) ! I, J, K of type integer.
```

By convention, such an object is called a substring rather than an array element.

Notice that it is the array *part-name* that the subscript list qualifies.  It is not permitted to apply such a subscript list to an array designator unless it applies to the last *part-name*.  An array section, a function reference, an array expression in parentheses, or a scalar component of an array structure must not be qualified by a subscript list.

## 6.10  Array sections

Array sections were introduced in Section 2.10 and provide a convenient way to access a regular subarray such as a row or a column of a rank-two array:

```
A(I, 1:N)   ! Elements 1 to N of row I
A(1:M, J)   ! Elements 1 to M of column J
```

For simplicity of description, we did not explain that one or both bounds may be omitted when the corresponding bound of the array itself is wanted, and that a stride other than one may be specified:

```
A(I, :)      ! The whole of row I
A(I, 1:N:3)  ! Elements 1, 4, ... of row I
```

Another form of section subscript is a rank-one integer expression.  All the elements of the expression must be defined with values that lie within the bounds of the parent array's subscript.  For example,

```
V( (/ 1, 7, 3, 2 /) )
```

is a section with elements V(1), V(7), V(3), and V(2), in this order.  Such a subscript is called a *vector subscript*.  If there are any repetitions in the values of the elements of a vector subscript, the section is called a *many-one section* because more than one element of the section is mapped onto a single array element. For example

```
V( (/ 1, 7, 3, 7 /) )
```

has elements 2 and 4 mapped onto V(7). A many-one section must not appear on the left of an assignment statement because there would be several possible values for a single element. For instance, the statement

```
V( (/ 1, 7, 3, 7 /) ) = (/ 1, 2, 3, 4 /)     ! Illegal
```

is not allowed because the values 2 and 4 cannot both be stored in V(7). The extent is zero if the vector subscript has zero size.

When an array section with a vector subscript is an actual argument, it is regarded as an expression and the corresponding dummy argument must not be defined or redefined and must not have intent OUT or INOUT. We expect compilers to make a copy as a temporary regular array on entry but to perform no copy back on return. Also, an array section with a vector subscript is not permitted to be a pointer target, since allowing them would seriously complicate the mechanism that compilers would otherwise have to establish for array pointers. For similar reasons, such an array section is not permitted to be an internal file (Section 9.6).

The general form of an array section is

*part-ref* [*%part-ref*]... [(*substring-range*)]

where *part-ref* now has the form

*part-name* [(*section-subscript-list*)]

where the number of section subscripts in each list must be equal to the rank of the array or array component. Each *section-subscript* is either a *subscript* (Section 6.9), a rank-one integer expression (vector subscript), or a *subscript-triplet* of the form

[*lower*] : [*upper*] [ : *stride*]

where *lower*, *upper*, and *stride* are scalar integer expressions. If *lower* is omitted, the default value is the lower bound for this subscript of the array. If *upper* is omitted, the default value is the upper bound for this subscript of the array. If *stride* is omitted, the default value is one. The stride may be negative so that it is possible to take, for example, the elements of a row in reverse order by specifying a section such as

```
A(I, 10:1:-1)
```

The extent is zero if *stride* > 0 and *lower* > *upper*, or if *stride* < 0 and *lower* < *upper*. The value of *stride* must not be zero.

Normally, we expect the values of both *lower* and *upper* to be within the bounds of the corresponding array subscript. However, all that is required is that each value actually used to select an element is within the bounds.

The *subscript-triplet* specifies a sequence of values, subscript *lower*, *lower+stride*, *lower+2\*stride*,... going as far as possible without going beyond *upper* (above it when *stride* > 0 or below it when *stride* < 0). The length of the sequence for the *i*-th *subscript-triplet* determines the *i*-th extent of the array that is formed.

The rank of a *part-ref* is the number of vector subscripts and subscript triplets that it contains. Exactly one must have non-zero rank and it determines the rank of the section. Its extents determine the shape. If any extent is zero, the section has size zero.

If an array section is of type character, it may be qualified by a substring reference. By convention, the resulting object is called a section rather than a substring. It is formed from the unqualified section by taking the specified substring of each element. Note that if C is a rank-one array,

```
C(I:J)
```

is the section formed from elements I to J; if substrings of all the array elements are wanted, we may write

```
C(:)(K:L)
```

## 6.11  Structure components that are arrays

In Section 6.9, we discussed structure components of the form

*structure % array-component-name*

which are arrays from which array elements and array sections may be selected. A structure component may also have the form

*array % scalar-component-name*

and this, too, is an array. It may be used as a whole in expressions and assignments and may be passed as an actual argument to a procedure, but it may not be qualified by (*subscript-list*) or (*section-subscript-list*). Using the example of Section 6.9,

```
TAR % U
```

is such an array and

```
TAR % U (1, 2, 3) ! Not permitted
```

is not allowed, though the same required element can be written

```
TAR (1, 2, 3) % U
```

The general form of a structure component is

*parent % component-name*

where *parent* is any object of derived type (it may itself be a structure component and it may be an array element or section). This is scalar if both the
parent and its component are scalar. It is an array if the parent is scalar and
its component is an array, or vice versa. It is a pointer if the component is
declared with the POINTER attribute in the type definition.

It is not permissible for both to be arrays, as in

```
TAR % DU    ! Illegal
```

The reason for this not being permitted is that if TAR%DU were an array,
element (1,2,3,4) would correspond to TAR(2,3,4)%DU(1), which would be
too confusing a notation. Additionally, if the parent is an array, the component must not have the POINTER attribute (arrays of pointers are very different from ordinary arrays, so array syntax is not allowed for them).

## 6.12  Pointers as aliases

If an array section without vector subscripts, such as

```
TABLE(M:N, P:Q)
```

is wanted frequently while the integer variables M, N, P, and Q do not change
their values, it is convenient to be able to refer to the section as a named array
such as

```
WINDOW
```

Such a facility is provided in Fortran 90 by pointer arrays and the pointer assignment statement. Here WINDOW would be declared thus

```
REAL, DIMENSION(:, :), POINTER :: WINDOW
```

and associated with TABLE by the execution of the statement

```
WINDOW  => TABLE(M:N, P:Q)
```

If, later on, the size of the window needs to be changed, all that is needed is another pointer assignment statement.   Note, however, that the subscript bounds for WINDOW in this example are (1:N–M+1, 1:Q–P+1) since they are as provided by the functions LBOUND and UBOUND (Section 8.12.2).

The facility provides a mechanism for subscripting or sectioning arrays such as

```
TAR % U
```

where TAR is an array and U is a scalar component, discussed in the previous section.  Here we may perform the pointer association

```
TARU => TAR % U
```

if TARU is a rank-three pointer array of the appropriate type.  Subscripting as in

```
TARU(1, 2, 3)
```

is then permissable.  Here the subscript bounds for TARU will be those of TAR.

## 6.13  Array constructors

The syntax that we introduced in Section 2.10 for array constants may be used to construct more general rank-one arrays.   The general form of an *array-constructor* is

(/ *array-constructor-value-list* /)

where each *array-constructor-value* is one of

*expr*

or
  *constructor-implied-do*

The array thus constructed is of rank one with its sequence of elements formed from the sequence of scalar expressions and elements of the array expressions in array-element order. A *constructor-implied-do* has the form

  ( *array-constructor-value-list, variable = expr1, expr2, [,expr3 ]* )

where *variable* is a named integer scalar variable, and *expr1, expr2,* and *expr3* are scalar integer expressions. Its interpretation is as if the *array-constructor-value-list* had been written

  MAX ( (*expr2 – expr1 + expr3*)/*expr3*, 0 )

times, with *variable* replaced by *expr1, expr1 + expr3,...,* as for the DO construct (Section 4.5). A simple example is

  (/ (I,I=1,10) /)

which is equal to

  (/ 1, 2, 3, 4, 5, 6, 7, 8, 9, 10 /)

Note that the syntax permits nesting of one *constructor-implied-do* inside another, as in the example

  (/ ((I,I=1,3), J=1,3) /)

which is equal to

  (/ 1, 2, 3, 1, 2, 3, 1, 2, 3 /)

The sequence may be empty, in which case a zero-sized array is constructed. The scope of the *variable* is the *constructor-implied-do*. Other statements, or even other parts of the array constructor, may refer to another variable having the same name. The value of the other variable is unaffected by execution of the array constructor and is available except within the *constructor-implied-do*.

  The type and type parameters of an array constructor are those of the first expression, and each expression must have the same type and type parameters. If every expression is a constant expression, the array constructor is a constant expression.

An array of rank greater than one may be constructed from an array constructor by using the intrinsic function RESHAPE (Section 8.13.3). For example,

```
RESHAPE( SOURCE = (/ 1,2,3,4,5,6 /), SHAPE = (/ 2,3 /) )
```

has the value

```
1 3 5
2 4 6
```

## 6.14  Mask arrays

Logical arrays are needed for masking in WHERE statements and constructs (Section 6.8), and they play a similar role in many of the array intrinsic functions (Chapter 8).  Often, such arrays are large, and there may be a worthwhile storage gain from using nondefault logical types, if available.  For example, some processors may use bytes to store elements of LOGICAL(KIND=1) arrays, and bits to store elements of LOGICAL(KIND=0) arrays.  Unfortunately, there is no *portable* facility to specify such arrays, since there is no intrinsic function comparable to SELECTED_INT_KIND and SELECTED_REAL_KIND.

## 6.15  Summary

We have explained that arrays may have zero size and that no special rules are needed for them.  A dummy array may assume its shape from the corresponding actual argument.  Storage for an array may be allocated automatically on entry to a procedure and automatically deallocated on return, or the allocation may be controlled in detail by the program.  Functions may be array-valued either through the mechanism of an elemental reference that performs the same calculation for each array element, or through the truly array-valued function.  Array assignments may be masked through the use of the WHERE statement and block.  Structure components may be arrays if the parent is an array or the component is an array, but not both.  A subarray may either be formulated directly as an array section, or indirectly by using pointer assignment to associate it with an pointer array.  An array may be constructed from a sequence of expressions.  A logical array may be used as a mask.

Basically, the whole of the contents of this chapter represents new features, and is a hallmark of the new standard.  The intrinsic functions are an important part of the array features of Fortran 90 and will be described in Chapter 8.

## Exercises

**1.** Given the array declaration

```
REAL, DIMENSION(50,20) :: A
```

write array sections representing

  i) the first row of A;

  ii) the last column of A;

  iii) every second element in each row and column;

  iv) as for (iii) in reverse order in both dimensions;

  v) a zero-sized array.

**2.** Write a WHERE statement to double the value of all the positive elements of an array Z.

**3.** Write an array declaration for an array J which is to be completely defined by the statement

```
J = (/ (3, 5, I=1,5), 5,5,5, (I, I = 5,3,-1 ) /)
```

**4.** Classify the following arrays:

```
SUBROUTINE EXAMPLE(N, A, B)
    REAL, DIMENSION(N, 10) :: W
    REAL A(:), B(0:)
    REAL, POINTER         :: D(:, :)
```

**5.** Write a declaration and a pointer assignment statement suitable to reference as an array all the third components of DU in the elements of the array TAR having all three subscript values even (Section 6.9).

**6.** Given the array declarations

```
INTEGER, DIMENSION(100, 100)     :: L, M, N
INTEGER, DIMENSION(:, :), POINTER :: LL, MM, NN
```

rewrite the statements

```
L(J:K+1, J-1:K) = L(J:K+1, J-1:K) + L(J:K+1, J-1:K)
L(J:K+1, J-1:K) = M(J:K+1, J-1:K) + N(J:K+1, J-1:K) + N(J:K+1, J:K+1)
```

as they could appear following execution of the statements

```
LL => L(J:K+1, J-1:K)
MM => M(J:K+1, J-1:K)
NN => N(J:K+1, J-1:K)
```

**7.** Complete Exercise 1 of Chapter 4 using array syntax instead of DO constructs.

**8.** Write a module to maintain a data structure consisting of a linked list of integers, with the ability to add and delete members of the list, efficiently.

**9.** Write a module that contains Figure 31 (Section 6.6) as a module procedure and supports the defined operations and assignments that it contains.

# 7. SPECIFICATION STATEMENTS

## 7.1 Introduction

In the preceding chapters we have learnt the elements of the Fortran 90 language, how they may be combined into expressions and assignments, how we may control the logic flow of a program, how to divide a program into manageable parts, and have considered how arrays may be processed. We have seen that this knowledge is sufficient to write programs, when combined with a rudimentary PRINT statement and with the END statement.

Already in Chapters 2 to 6, we met some specification statements when declaring the type and other properties of data objects, but to ease the reader's task we did not explain all the available options. In this chapter we fill this gap, describing the full syntax of the specification statements and associated details. To begin with, however, it is necessary to recall the place of specification statements in a programming language. A program is processed by a computer in (usually) three stages. In the first stage, *compilation*, the source code (text) of the program is read and processed by a program known as a *compiler* which analyses it, and generates a file containing *object code*. Each program unit of the complete program is usually processed separately. The object code is a translation of the source code into a form which can be understood by the computer hardware, and contains the precise instructions as to what operations the computer is to perform. In the second stage of processing, the object code is placed in the relevant part of the computer's storage system by a program often known as a loader which prepares it for the next stage; during this second stage, the separate program units are linked to one another, that is joined to form a complete executable program. The third stage consists of the *execution*, whereby the coded instructions are performed and the results of the computations made available.

During the first stage, the compiler requires information about the entities involved. This information is provided at the beginning of each program unit or subprogram by specification statements, whose description is the subject of this chapter.

## 7.2  Implicit typing

Many programming languages require that all typed entities have their types specified explicitly.  Any data entity that is encountered in an executable statement without its type having been declared will cause the compiler to indicate an error.  This, and a prohibition on mixing types, is known as *strong typing*.  In the case of Fortran 90, an entity appearing in the code without having been explicitly typed is normally *implicitly* typed, being assigned a type according to its initial letter.  The default in a program unit is that entities whose names begin with one of the letters I, J,..., N are of type default integer, and variables beginning with the letters A, B,..., H or O, P,..., Z are of type default real.  This absence of strong typing can lead to program errors; for instance, if a variable name is misspelt, the misspelt name will give rise to a separate variable.  For this reason, we recommend that implicit typing be avoided.

Implicit typing does not apply to an entity accessed by use or host association because its type is the same as in the module or the host.

If a different rule for implicit typing is desired in a given scoping unit, the IMPLICIT statement may be employed.  For no implicit typing whatsoever, the statement

```
IMPLICIT NONE
```

is available (our recommendation), and for changing the mapping between the letters and the types, statements such as

```
IMPLICIT INTEGER (A-H)
IMPLICIT REAL(SELECTED_REAL_KIND(10)) (R,S)
IMPLICIT TYPE(ENTRY) (U,X-Z)
```

are available.  The letters are specified as a list in which a set of adjacent letters in the alphabet may be abbreviated, as in A–H.  No letter may appear twice in the implicit statements of a scoping unit and if there is an IMPLICIT NONE statement, there must be no other implicit statements in the scoping unit.  For a letter not included in the implicit statements, the mapping between the letter and a type is the default mapping.

In the case of a nested set of scoping units, such as a module containing a module subprogram, the default mapping for each letter in an inner scoping unit is the mapping for the letter in the immediate host.

Figure 34 provides a comprehensive illustration of the rules of implicit typing.

```
MODULE EXAMPLE_MOD
   IMPLICIT NONE
   :
   INTERFACE
      FUNCTION FUN(I)      ! All data entities must
         INTEGER FUN, I    ! be declared explicitly.
      END FUNCTION FUN
   END INTERFACE
CONTAINS
   FUNCTION JFUN(J)        ! All data entities must
      INTEGER JFUN, J      ! be declared explicitly.
      :
   END FUNCTION
END MODULE
SUBROUTINE SUB
   IMPLICIT COMPLEX (C)
   C = (3.0,2.0)           ! C is implicitly declared COMPLEX.
   :
CONTAINS
   SUBROUTINE SUB1
      IMPLICIT INTEGER (A,C)
      C = (0.0,0.0)        ! C is host associated and of type COMPLEX
      Z = 1.0              ! Z is implicitly declared REAL.
      A = 2                ! A is implicitly declared INTEGER.
      CC = 1.0             ! CC is implicitly declared INTEGER.
      :
   END SUBROUTINE SUB1
   SUBROUTINE SUB2
      Z = 2.0              ! Z is implicitly declared REAL and is
                           ! different from the variable Z in SUB1.
      :
   END SUBROUTINE SUB2
   SUBROUTINE SUB3
   USE EXAMPLE_MOD         ! Access the integer function FUN
      Q = FUN(K)           ! Q is implicitly declared REAL and
                           ! K is implicitly declared INTEGER.
      :
   END SUBROUTINE SUB3
END SUBROUTINE SUB
```

Figure 34.

The general form of the IMPLICIT statement is

**IMPLICIT NONE**

or

**IMPLICIT** *type* (*letter-spec-list*) [, *type* (*letter-spec-list*) ]...

where *type* specifies the type and type parameters (Section 7.13) and each *letter-spec* is

> letter [— letter ]

The IMPLICIT statement may be used for a derived type. For example, given access to the type

```
TYPE POSN
   REAL X, Y
   INTEGER Z
END TYPE POSN
```

and given the statement

```
IMPLICIT TYPE(POSN) (A,B), INTEGER (C-Z)
```

variables beginning with the letters A and B are implicitly typed POSN and variables beginning with the letters C, D,..., Z are implicitly typed INTEGER.

An IMPLICIT NONE statement may be preceded within a scoping unit only by USE (and FORMAT) statements, and other IMPLICIT statements may be preceded only by USE, PARAMETER, and FORMAT statements. We recommend that all IMPLICIT statements be at the start of the specifications, immediately following any USE statements.

## 7.3  Declaring entities of differing shapes

So far, we have used separate type declaration statements such as

```
INTEGER A, B
INTEGER, DIMENSION(10)  :: C, D
INTEGER, DIMENSION(8,7) :: E
```

to declare several entities of the same type but differing shapes. In fact, Fortran 90 permits the convenience of using a single statement. Whether or not there is an DIMENSION attribute present, arrays may be declared by placing the shape information after the name of the array:

```
INTEGER A, B, C(10), D(10), E(8, 7)
```

If the DIMENSION attribute is present, it provides a default shape for the entities that are not followed by their own shape information, and is ignored for those that are:

```
INTEGER, DIMENSION(10) :: C, D, E(8, 7)
```

Finally, if all the entities involved are arrays, they may be declared *without* type specifications in a DIMENSION statement:

```
DIMENSION I(10), B(50,50), C(N,M)   ! N and M are dummy
                                    ! scalar integer arguments
```

The general form is

DIMENSION *array-name(array-spec)* [*,array-name(array-spec)* ]...

Here, the type may either be specified in a type declaration statement such as

```
INTEGER I
```

that does not specify the dimension information, or be specified implicitly. Our view is that neither of these is sound practice: the type declaration statement looks like a declaration of a scalar and we explained in the previous section that we regard implicit typing as dangerous. Therefore, the use of the DIMENSION statement is not recommended.

## 7.4  Named constants and constant expressions

Inside a program, we often need to define a constant or set of constants. For instance, in a program requiring repeated use of the speed of light, we might use a real variable C that is given its value by the statement

```
C = 2.99792458
```

A danger in this practice is that the value of C may be overwritten inadvertently, for instance because another programmer re-uses C as a variable to contain a different quantity, failing to notice that the name is already in use.

Another situation which can arise is that the program contains specifications such as

```
REAL    X(10), Y(10), Z(10)
INTEGER MESH(10, 10), IPOINT(100)
```

where all the dimensions are 10 or $10^2$. Such specifications may be used extensively, and 10 may even appear as an explicit constant, say as a parameter in a DO-loop which processes these arrays:

```
DO I = 1, 10
```

Later, it may be realised that the value 20 rather than 10 is required, and the new value must be substituted everywhere the old one occurs, an error-prone undertaking.

Yet another case was met in Section 2.6, where a named constant was needed for the kind type parameter values.

In order to deal with all of these situations, Fortran 90 contains what are known as *named constants*. These may never appear on the left-hand side of an assignment statement, but may be used in expressions in any way in which a literal constant may be used. A type declaration statement may be used to specify such a constant:

```
REAL, PARAMETER :: C = 2.99792458
```

The value is protected, as C is now the name of a constant and may not be used as a variable name in the same scoping unit. Similarly, we may write

```
INTEGER, PARAMETER :: LENGTH = 10
REAL    X(LENGTH), Y(LENGTH), Z(LENGTH)
INTEGER MESH(LENGTH, LENGTH), IPOINT(LENGTH**2)
:
DO I = 1, LENGTH
```

which has the clear advantage that in order to change the value of 10 to 20 only a single line need be modified, and the new value is then correctly propagated.

In this example, the expression LENGTH**2 appeared in one of the array bound specifications. This is a particular example of a constant expression. A *constant expression* is an expression in which each operation is intrinsic, and each primary is

  i) a constant or a subobject of a constant where each subscript, section subscript, and substring bound is a constant expression,

 ii) an array constructor whose expressions (including bounds and strides) have primaries that are constant expressions or implied-DO variables,

iii) a structure constructor whose components are constant expressions,

 iv) an elemental intrinsic function reference whose arguments are constant expressions,

  v) a transformational intrinsic function reference whose arguments are constant expressions,

vi) a reference to an inquiry function (Section 8.1.2) other than PRESENT, ASSOCIATED, or ALLOCATED, where each argument is either a constant expression or a variable whose type parameters or bounds inquired about are not assumed or defined by an ALLOCATE statement or a pointer assignment, or

vii) a constant expression enclosed in parentheses.

Because the values of named constants are expected to be evaluated at compile time, the expressions permitted for their definition are restricted in their form. An *initialization expression* is a constant expression in which

i) the exponentiation operator must have an integer power,

ii) an elemental intrinsic function must have arguments and results of type integer or character, and

iii) of the transformational functions, only REPEAT, TRIM, TRANSFER, and RESHAPE are permitted.

If an initialization expression invokes an inquiry function for a type parameter or an array bound of an object, the type parameter or array bound must be specified in a prior specification statement or to the left in the same specification statement.

In the definition of a named constant we may use any initialization expression, and the constant becomes defined with the value of the expression according to the rules of intrinsic assignment. This is illustrated by the example:

```
INTEGER, PARAMETER :: LENGTH=10, LONG=SELECTED_REAL_KIND(12)
REAL, PARAMETER    :: LSQ = LENGTH**2
```

Note from this example that it is possible in one statement to define several named constants, in this case two, separated by commas.

A named constant may be an array, as in the case

```
REAL, DIMENSION(3), PARAMETER :: FIELD = (/ 0., 10., 20. /)
```

and it may be of derived type, as in the case

```
TYPE(POSN), PARAMETER :: A = POSN(1.0,2.0,0)
```

for the type defined at the end of Section 7.2. Note that a subobject of a constant need not necessarily have a constant value. For example, if I is an integer variable, FIELD(I) may have the value 0., 10., or 20. Note also that a constant may not be a pointer, dummy argument, or function result, since these are always variables.

An alternative way to specify a named constant is by the PARAMETER statement. It has the general form

PARAMETER ( *named-constant-definition-list* )

where each *named-constant-definition* is

*constant-name = initialization-expression*

Each constant named must either have been typed in a previous type declaration statement in the scoping unit, or take its type from the first letter of its name according to the implicit typing rule of the scoping unit. In the case of implicit typing, the appearance of the named constant in a subsequent type declaration statement in the scoping unit must confirm the type and type parameters. Similarly, the shape must have been specified previously or be scalar. Each named constant in the list is defined with the value of the corresponding expression according to the rules of intrinsic assignment.

Any named constant used in an initialization expression must either be accessed from the host, or be accessed from a module, or be declared in a preceding statement, or be declared to the left of its use in the same statement. An example using implicit typing and a constant expression including a named constant that is defined in the same statement is

```
IMPLICIT INTEGER (A, P)
PARAMETER (APPLE = 3, PEAR = APPLE**2)
```

Finally, there is an important point concerning the definition of a scalar named constant of type CHARACTER. Its length may be specified as an asterisk and taken directly from its value, which obviates the need to count the length of character strings, and makes modifications to their definitions much easier. An example of this is

```
CHARACTER(LEN=*), PARAMETER :: STRING = 'No need to count'
```

The PARAMETER attribute is an important means whereby constants may be protected from overwriting, and programs modified in a safe way. It should be used for these purposes on every possible occasion. On the other hand, for exactly the same reasons as for DIMENSION, we recommend avoiding the PARAMETER statement.

## 7.5 Initial values for variables

A variable may be assigned an initial value in a type declaration statement, simply by following the name of the variable by an initialization expression (Section 7.4), as in the examples:

```
REAL              :: A = 0.0
REAL, DIMENSION(3) :: B = (/ 0.0, 1.2, 4.5 /)
```

The initial value is defined by the value of the corresponding expression according to the rules of intrinsic assignment. If any part of a variable is initialized, the variable automatically acquires the SAVE attribute (Section 7.9).

No variable that is given an initial value may be a dummy argument, accessed by use or host association, a pointer, an allocatable array, an automatic object, or a function result.

An alternative way to specify an initial value for such a variable is by the DATA statement. It has the general form

DATA *object-list/value-list/[[,] object-list/value-list]...*

where *object-list* is a list of variables and *implied-DO loops*; and *value-list* is a list of scalar constants and structure constructors. A simple example is

```
REAL    A, B, C
INTEGER I, J, K
DATA    A,B,C/1.,2.,3./, I,J,K/1,2,3/
```

in which a variable A acquires the initial value 1., B the value 2., etc. After any array or array section in *object-list* has been expanded into a sequence of scalar elements in array element order, there must be as many constants in each *value-list* as scalar elements in the corresponding *object-list*. Each scalar element is assigned the corresponding scalar constant.

Constants which repeat may be written once and combined with a scalar integer *repeat count* which may be a named or literal constant:

```
DATA I,J,K/3*0/
```

The value of the repeat count must be positive or zero. As an example consider the statement

```
DATA R(1:LENGTH)/LENGTH*0./
```

where R is a real array and LENGTH is a named constant which might take the value zero.

Arrays may be initialized in three different ways: as a whole, by element, or by an implied-DO loop. These three ways are shown below for an array declared by

```
REAL A(5, 5)
```

Firstly, for the whole array, the statement

```
DATA A/25*1.0/
```

sets each element of A to 1.0.

Secondly, individual elements and sections of A may be initialized, as in

```
DATA A(1,1), A(3,1), A(1,2), A(3,3) /2*1.0, 2*2.0/
DATA A(2:5,4) /4*1.0/
```

in each of which only the four specified elements and the section are initialized.

When the elements to be selected fall into a pattern which can be represented by DO-loop indices, it is possible to write DATA statements a third way, like

```
DATA (( A(I,J), I=1,5,2), J=1,5) /15*0./
```

The general form of an implied-DO loop is

( *dlist, do-var = expr, expr, [expr ]*)

where *dlist* is a list of array elements and implied-DO loops, *do-var* is a named integer scalar variable, and each *expr* is a scalar integer expression. It is interpreted as for a DO construct (Section 4.5) except that the DO variable has the scope of the implied-DO as in an array constructor (Section 6.13). A variable in an *expr* must be a *do-var* of an outer implied-DO:

```
INTEGER J, K
INTEGER, PARAMETER :: L=5, L2=((L+1)/2)**2
REAL A(L,L)
DATA ((A(J,K), K=1,J), J=1,L,2) / L2 * 1.0 /
```

This example sets to 1.0 the first element of the first row of A, the first three elements of the third row, and all the elements of the last row, as shown in Figure 35.

| 1.0 | . | . | . | . |
|-----|---|---|---|---|
| . | . | . | . | . |
| 1.0 | 1.0 | 1.0 | . | . |
| . | . | . | . | . |
| 1.0 | 1.0 | 1.0 | 1.0 | 1.0 |

Figure 35. Result of an implied-DO loop in a DATA statement.

The only variables permitted in subscript expressions in DATA statements are DO indices of the same or an outer level loop.

An object of derived type may appear in a DATA statement. In this case, the corresponding value must be a structure constructor having a constant expression for each component. Using the type definition of POSN in Section 7.2, we can write

```
TYPE(POSN) :: POSITION1, POSITION2
DATA POSITION1 /POSN(2., 3., 0)/, POSITION2%Z /4/
```

In the examples given so far, the types and type parameters of the constants in a *value-list* have always been the same as the type of the variables in the *object-list*. This need not be the case, but they must be compatible for intrinsic assignment since the entity is initialized following the rules for intrinsic assignment. It is thus possible to write statements such as

```
DATA Q/1/, I/3.1/, B/(0.,1.)/
```

(where B and Q are real and I is integer). Integer values may be binary, octal, or hexadecimal constants (Section 2.6.1).

Each variable must either have been typed in a previous type declaration statement in the scoping unit, or its type is that associated with the first letter of its name according to the implicit typing rules of the scoping unit. In the case of implicit typing, the appearance of the name of the variable in a subsequent type declaration statement in the scoping unit must confirm the type and type parameters. Similarly, any array variable must have previously been declared as such.

No variable or part of a variable may be initialized more than once in a scoping unit.

As for DIMENSION and PARAMETER, we recommend using the type declaration statement rather than the DATA statement, but the DATA statement *must* be employed when only part of a variable is to be initialized.

## 7.6  PUBLIC and PRIVATE attributes

Modules (Section 5.5) permit specifications to be 'packaged' into a form that allows them to be accessed elsewhere in the program.  So far, we have assumed that all the entities in the module are to be accessible, that is have the PUBLIC attribute, but sometimes it is desirable to limit the access (for example, to prevent accidental changes).  In cases where entities are not to be accessible outside their own module, they may be given the PRIVATE attribute.  These two attributes may be specified with the PUBLIC and PRIVATE attributes on type declaration statements in the module, as in

```
REAL, PUBLIC    :: X, Y, Z
INTEGER, PRIVATE :: U, V, W
```

or in PUBLIC and PRIVATE statements, as in

```
PUBLIC  :: X, Y, Z, OPERATOR(.ADD.)
PRIVATE :: U, V, W, ASSIGNMENT(=), OPERATOR(*)
```

which have the general forms

PUBLIC [ [ :: ] *access-id-list*]
PRIVATE [ [ :: ] *access-id-list*]

where *access-id* is a name or a *generic-spec* (Section 5.18).

Note that if a procedure has a generic identifier, the accessibility of its specific name is independent of the accessibility of its generic identifier.  One may be PUBLIC while the other is PRIVATE, which means that it is accessible only by its specific name or only by its generic identifier.

If a PUBLIC or PRIVATE statement has no list of entities, it confirms or resets the default.  Thus the statement

```
PUBLIC
```

confirms PUBLIC as the default value, and the statement

```
PRIVATE
```

sets the default value for the module to PRIVATE accessibility. For example,

```
PRIVATE
PUBLIC :: MEANS
```

gives the entity MEANS the PUBLIC attribute whilst all others are PRIVATE. There may be at most one accessibility statement without a list in a scoping unit.

The entities which may be specified by name in PUBLIC or PRIVATE lists are named variables, non-intrinsic procedures, generic procedures, derived types, named constants, and namelist groups. An entity with a type must not have the PUBLIC attribute if its type has the PRIVATE attribute (because there is virtually nothing one can do with such an entity without access to its type). If a module procedure has a dummy argument or function result of PRIVATE type, the procedure must be given the attribute PRIVATE and must not have a generic identifier that is PUBLIC. The PUBLIC and PRIVATE attributes may appear only in the specifications of a module. The use of the PRIVATE statement for components of derived types in this context will be described in Section 7.11.

## 7.7  POINTER, TARGET, & ALLOCATABLE statements

For the sake of regularity in the language, there are statements for specifying the POINTER, TARGET, and ALLOCATABLE attributes of entities. They take the forms:

> POINTER [::] *object-name*[(*array-spec*)]
>      [,*object-name* [(*array-spec*)]]...

> TARGET [::] *object-name*[(*array-spec*)]
>      [,*object-name* [(*array-spec*)]]...

and

> ALLOCATABLE [::] *array-name*[(*array-spec*)]
>      [,*array-name* [(*array-spec*)]]...

as in

```
REAL A, SON, Y
ALLOCATABLE :: A(:,:)
POINTER :: SON
TARGET :: A, Y(10)
```

We believe that it is much clearer to specify these attributes on the type declaration statements, and therefore do not use these forms.

## 7.8  INTENT and OPTIONAL statements

The INTENT attribute (Section 5.9) for a dummy argument that is not a dummy procedure or pointer may be specified in a type declaration statement or in an INTENT statement of the form

INTENT( *inout* ) [::] *dummy-argument-name-list*

where *inout* is IN, OUT, or INOUT.

Examples are

```
INTENT(IN)  :: A, B, C
INTENT(OUT) :: X, Y, Z
```

The meaning of the INTENT attribute was explained in Section 5.9.   It may not appear in a specification statement of a main program.
The OPTIONAL attribute (Section 5.13) for a dummy argument may be specified in a type declaration statement or in an OPTIONAL statement of the form

OPTIONAL [::] *dummy-argument-name-list*

An example is

```
OPTIONAL :: A, B, C
```

The meaning of the OPTIONAL attribute was explained in Section 5.13. It may not appear in a specification statement of a main program; it is the only attribute which may be specified for a dummy argument which is a procedure.

## 7.9  The SAVE attribute

Let us suppose that we wish to retain the value of a local variable in a subprogram, for example to count the number of times the subprogram is entered. We might write a section of code as in Figure 36.  In this example, the local variables, A and COUNTER, are initialized to zero, and it is assumed that their current values are available each time the subroutine is called.  This is not necessarily the case.  Fortran 90 allows the computer system being used to 'forget' a new value, the variable becoming undefined on each return unless it has the SAVE attribute. In the Figure 36, it is sufficient to change the declaration of A to

```
REAL, SAVE :: A
```

to be sure that its value is always retained between calls. This may be done for COUNTER, too, but is not necessary as all variables with initial values acquire the SAVE attribute automatically (Section 7.5).

```
SUBROUTINE ANY(X)
   REAL A, X
   INTEGER :: COUNTER = 0 ! Initialize the counter
   :
   COUNTER = COUNTER + 1
   IF (COUNTER.EQ.1) THEN
      A = 0.
   ELSE
      A = A + X
   END IF
   :
```

Figure 36.

A similar situation arises with the use of variables in modules (Section 5.5). On return from a subprogram that accesses a variable in a module, the variable becomes undefined unless the main program accesses it, another subprogram in execution accesses it, or it has the SAVE attribute.

The SAVE attribute must not be specified for a dummy argument, a function result, or an automatic object (Section 6.4). It may be specified for a pointer, in which case the pointer association status is saved. It may be specified for an allocatable array, in which case the allocation status and value are saved.

An alternative to specifying the SAVE attribute on a type declaration statement is the SAVE statement:

SAVE [ [::] *variable-name-list* ]

A SAVE statement with no list is equivalent to a list containing all possible names, and in this case the scoping unit must contain no other SAVE statements and no SAVE attributes in type declaration statements.

The SAVE statement or SAVE attribute may appear in the declaration statements in a main program but has no effect.

## 7.10  The USE statement

In Section 5.5, we introduced the USE statement in its simplest form

USE *module-name*

which provides access to all the public named data objects, derived types, interface blocks, procedures, generic identifiers, and namelist groups in the module named. USE statements must precede other specification statements in a scoping unit.

If access is needed to two or more modules that have been written independently, the same name may be used in more than one module. This is the main reason for permitting accessed entities to be renamed by the USE statement. Renaming is also available to resolve a name clash between a local entity and an entity accessed from a module, though our preference is to use a text editor or other tool to change the local name. With renaming, the USE statement has the form

USE *module-name, rename-list*

where each *rename* has the form

*local-name => use-name*

and refers to a public entity in the module that is to be accessed by a different local name.

As an example,

```
USE STATS_LIB, SPROD => PROD
USE MATHS_LIB
```

makes all the public entities in both STATS_LIB and MATHS_LIB accessible. If MATHS_LIB contains an entity called PROD, it is accessible by its own name while the entity PROD of STATS_LIB is accessible by the name SPROD.

Renaming is not needed if there is a name clash between two entities that are not required. A name clash is permitted if there is no reference to the name in the scoping unit.

It is also permissible for a generic name that is required. Here, all the generic interfaces accessed by this name are treated as a single concatenated interface block. This is true also for defined operators and assignments, where no renaming facility is available. In all these cases, any two procedures having the same generic identifier must differ as explained in Section 5.18. We imagine that this will usually be exactly what is needed. For example, we might access modules for interval arithmetic and matrix arithmetic, both needing the functions SQRT, SIN, etc., the operators +, −, etc., and assignment, but for different types.

For cases where only a subset of the names of a module is needed, the ONLY option is available, having the form

USE *module-name*, ONLY : [*only-list*]

where each *only* has the form

*access-id*

or

[*local-name* => ] *use-name*

where each *access-id* is a public entity in the module, and is either a *use-name* or a *generic-spec* (Section 5.18). This provides access to an entity in a module only if the entity is public and is specified as a *use-name* or *access-id*. Where a *use-name* is preceded by a *local-name*, the entity is known locally by the *local-name*. An example of such a statement is

```
USE STATS_LIB, ONLY : SPROD => PROD, MULT
```

which provides access to PROD by the local name SPROD and to MULT by its own name.

We would recommend that only one USE statement for a given module be placed in a scoping unit, but more are allowed. If there is a USE statement without an ONLY qualifier, all public entities in the module are accessible and the *rename-list*s and *only-list*s are interpreted as if concatenated into a single *rename-list*. If all the statements have the ONLY qualification, only those entities named in one or more of the *only-list*s are accessible, that is all the *only-list*s are interpreted as if concatenated into a single *only-list*.

The form

USE *module-name*, ONLY :

might appear redundant. It is provided for the situation where a scoping unit calls a set of procedures that communicate with each other through shared data in a module. It ensures that the data are available throughout the execution of the scoping unit.

When a module contains USE statements, the entities accessed are treated as entities in the module. They may be given the PRIVATE or PUBLIC attribute explicitly or through the default rule in effect in the module. No other attribute of an accessed entity may be specified afresh, but the entity may be included in one or more namelist groups (Section 7.15).

An entity may be accessed by more than one local name. This is illustrated in Figure 37, where module B accesses S of module A by the local name BS; if a subprogram such as C accesses both A and B, it will access S by both its original name and by the name BS. Example Figure 37. also illustrates that an entity may be accessed by the same name by more than one route (see variable T).

```
MODULE A
   REAL S, T
   :
END MODULE A
MODULE B
   USE A, BS => S
   :
END MODULE B
SUBROUTINE C
   USE A
   USE B
   :
END SUBROUTINE
```

Figure 37.

A more direct way for an entity to be accessed by more than one local name is for it to appear more than once as a *use-name*. This is not a practice that we recommend.

Of course, all the local names of entities accessed from modules must differ from each other and from names of local entities. If a local entity is accidently given the same name as an accessible entity from a module, this will be noticed at compile time if the local entity is declared explicitly. However, if the local entity is implicitly typed (Section 7.2) and appears in no specification statements, then each appearance of the name will be taken, incorrectly, as a reference to the accessed variable. To avoid this, we recommend the use of

```
IMPLICIT NONE
```

in a scoping unit containing one or more USE statements.

## 7.11  Derived-type definitions

When derived types were introduced in Section 2.9, some simple example definitions were given, but the full generality was not included. The general form (apart from redundant features, see Sections 11.2 and 11.4.4) is

TYPE [[, *access*]:: ] *type-name*
  [ PRIVATE ]
  *component-def-stmt*
  [*component-def-stmt*]...
END TYPE [ *type-name* ]

Each *component-def-stmt* has the form

*type* [ [ ,*component-attr-list* ] :: ]*component-decl-list*

where *type* specifies the type and type parameters (Section 7.13), each *component-attr* is either POINTER or DIMENSION(*extent-list*), and each *component-decl* is

*component-name* [ (*extent-list*) ][ *\*char-len* ]

The meaning of *\*char-len* is explained in Section 7.13. If the *type* is a derived type and the POINTER attribute is not specified, the type must be previously defined in the host scoping unit or accessible there by use or host association. If the POINTER attribute is specified, the type may also be the one being defined (for example, the type ENTRY of Section 2.13), or one defined later in the scoping unit.

A *type-name* must not be the same as the name of any intrinsic type or any other accessible derived type.

The bounds of an array component are declared by an *extent-list* where each *extent* is

:

for a pointer component or

[*lower-bound*:] *upper-bound*

for a non-pointer component, and *lower-bound* and *upper-bound* are constant expressions that are restricted to specification expressions (Section 7.14). If there is an *extent-list* attached to the *component-name*, this defines the bounds. If an DIMENSION attribute is present in the statement, its *extent-list* applies to any component in the statement without its own *extent-list*.

Only if the host scoping unit is a module may the *access* qualifier or PRIVATE statement appear. The *access* qualifier on a type statement may be PUBLIC or PRIVATE and specifies the accessibility of the type. If it is PRIVATE, then the type name, the structure constructor for the type, any entity of the type, and any procedure with a dummy argument or function result of the type are all inaccessible outside the host module. The accessibility may also be specified in a PRIVATE or PUBLIC statement in the host. In the absence of both of these, the type takes the default accessibility of the host module. If a PRIVATE statement appears for a type with PUBLIC accessibility, the components of the type are inaccessible in any scoping unit accessing the host module, so that neither component selection nor structure construction are available there. Also, if any component is of a derived type that is PRIVATE, the type being defined must be PRIVATE or have PRIVATE components.

We can thus distinguish three levels of access:

 i) all PUBLIC, where the type and all its components are accessible;

 ii) a PUBLIC type with PRIVATE components, where the type is accessible but its components are hidden;

 iii) all PRIVATE, where both the type and its components are used only within the host module, and are hidden to an accessing procedure.

Case ii) has, where appropriate, the advantage of enabling changes to be made to the type without in any way affecting the code in the accessing procedure. Case iii) offers this advantage and has the additional merit of not cluttering the name space of the accessing procedure. The use of PRIVATE accessibility for the components or for the whole type is thus recommended whenever possible.

We note that even if two derived-type definitions are identical in every respect except their names, that entities of those two types are *not* equivalent and are regarded as being of different types. Even if the names, too, are identical, the types are different (unless they have the SEQUENCE attribute, a feature that we do not recommend and whose description is left to Section 11.2). If a type is needed in more than one program unit, the definition should be placed in a module and accessed by a USE statement wherever it is needed. Having a single definition is far less prone to errors.

## 7.12  The type declaration statement

We have already met many simple examples of the declarations of named entities by INTEGER, REAL, COMPLEX, LOGICAL, CHARACTER, and TYPE( *type-name*) statements. The general form is

   *type* [ [ , *attribute*]... ::  ] *entity-list*

where *type* specifies the type and type parameters (Section 7.13), *attribute* is one of the following

```
PARAMETER
PUBLIC
PRIVATE
POINTER
TARGET
ALLOCATABLE
DIMENSION(extent-list)
INTENT(inout)
OPTIONAL
SAVE
EXTERNAL
INTRINSIC
```

and each *entity* is

   *object-name* [ *(extent-list)* ] [ **char-len*] [ = *initialization-expr* ]
or
   *function-name* [ *(extent-list)* ] [ **char-len*]

The meaning of **char-len* is explained in Section 7.13.

   No attribute may appear more than once in a given type declaration statement. The double colon :: must appear wherever =*initialization-expr* appears or an attribute is specified. If the statement specifies a PARAMETER attribute, =*initialization-expr* must appear.

   A function name must be the name of an external, dummy, or intrinsic function.

   If a POINTER attribute is specified, the INTENT, EXTERNAL, and INTRINSIC attributes must not be specified. The TARGET and PARAMETER attributes may not be specified for the same entity, and the POINTER and ALLOCATABLE attributes may not be specified for the same array. If the TARGET attribute is specified, neither the EXTERNAL nor the INTRINSIC attribute may also be specified.

   A dummy argument or function result must not be specified with the ALLOCATABLE, PARAMETER, or SAVE attribute.

   The INTENT and OPTIONAL attributes may be specified only for dummy arguments.

   Specifying the EXTERNAL attribute is an alternative way to declare an external function to the EXTERNAL statement (Section 5.11), and specifying the INTRINSIC attribute is an alternative to the INTRINSIC statement for declaring functions to be intrinsic (Section 8.1.3). These two attributes are mutually exclusive.

   Any of the attributes may also be specified in statements (such as SAVE) that list entities having the attribute. This leads to the possibility of an attribute being specified explicitly more than once for a given entity, but this is not permitted. Our recommendation is to avoid such statements because it is much clearer to have all the attributes for an entity collected in one place.

## 7.13  Type and type parameter specification

We have used *type* to represent one of the following

> INTEGER [( [KIND=] *kind-value*)]
> REAL  [( [KIND=] *kind-value*)]
> COMPLEX  [( [KIND=] *kind-value*)]
> CHARACTER [(*actual-parameter-list*)]
> LOGICAL [([KIND=] *kind-value*) ]
> TYPE ( *type-name* )

in the FUNCTION statement (Section 5.20), the IMPLICIT statement (Section 7.2), the component definition statement (Section 7.11), and the type declaration statement (Section 7.12). A *kind-value* must be an initialization expression (Section 7.4) and must never be negative.

For CHARACTER, each *actual-parameter* has the form

> [LEN=] *len-value*

or

> [KIND=] *kind-value*

and provides a value for one of the parameters. It is permissible to omit KIND= from a KIND *actual-parameter* only when LEN= is omitted and *len-value* is present, just as for an actual argument list (Section 5.13). Neither parameter may be specified more than once.

For a scalar named constant or for a dummy argument of a subprogram, a *len-value* may be specified as an asterisk, in which case the value is assumed from that of the constant itself or the associated actual argument.

A *len-value* that is not an asterisk must be a specification expression (Section 7.14). A negative value is allowed and declares character entities of zero length.

In addition, it is possible to attach an alternative form of *len-val* to individual entities in a type declaration statement using the syntax *entity*char-len*, where *char-len* is (*len-value*) or *len* and *len* is a scalar integer literal constant which specifies a length for the entity.

An example illustrating this form is

```
CHARACTER(LEN=8) WORD(4), POINT*1, TEXT(20)*4
```

Here, WORD, POINT and TEXT have character length 8, 1, and 4, respectively. Similarly, the alternative form may be used for individual components in a component definition statement.

## 7.14  Specification expressions

Nonconstant scalar integer expressions may be used to specify the array bounds and character lengths of data objects in a subprogram, and of function results. Such an expression may depend only on data values that are defined on entry to the subprogram. Any variable referenced must not have its type and type parameters specified later in the same sequence, unless they are those implied by the implicit typing rules.

Intrinsic function references are limited to:

i) an elemental function reference for which the arguments and results are of type integer or character,

ii) a reference to REPEAT, TRIM, TRANSFER, or RESHAPE for which the arguments are of type integer or character,

iii) a reference to SELECTED_INT_KIND or SELECTED_REAL_KIND,

iv) an inquiry function reference other than PRESENT, ASSOCIATED, or ALLOCATED, provided the quantity inquired about does depend on an allocation or on a pointer assignment.

Array constructors and derived-type constructors are permitted, but references to non-intrinsic procedures are not permitted. The expression may reference an inquiry function for an array bound or for a type parameter of an entity which either is accessed by use or host association, or is specified earlier in the same specification sequence, but not later in the sequence.[11] An element of an array specified in the same specification sequence can be referenced only if the bounds of the array are specified earlier in the sequence.[12] Such an expression is called a *specification expression*.

---

[11] This avoids such a case as

```
FUNCTION FUN(A)
REAL(KIND(A)) FUN
REAL(KIND(FUN)) A
```

[12] This avoids such a case as

```
FUNCTION FUN(I, J)
INTEGER I(J(1):J(2)), J(I(1):I(2))
```

An array whose bounds are declared using specification expressions is called an *explicit-shape array*.

The bounds and character lengths are not affected by any redefinitions or undefinitions of variables in the expressions during execution of the procedure.

## 7.15  The NAMELIST statement

It is sometimes convenient to gather a set of variables into a single group, in order to facilitate input/output (I/O) operations on the group as a whole. The actual use of such groups is explained in Section 9.10. The method by which a group is declared is via the NAMELIST statement which in its simple form has the syntax

NAMELIST *namelist-spec*

where *namelist-spec* is

*/namelist-group-name/ variable-name-list*

The *namelist-group-name* is the name given to the group for subsequent use in the I/O statements. A variable named in the list must not be a dummy array with non-constant bounds, a variable with assumed character length, an automatic object, an allocatable array, a pointer, or have a component that is a pointer at any depth of component selection. An example is

```
REAL BRUSHES(10)
NAMELIST /HOUSEHOLD_ITEMS/ CARPET, TV, BRUSHES
```

It is possible to declare several namelist groups in one statement, with the syntax

NAMELIST *namelist-spec* [ [,]*namelist-spec*]...

as in the example

```
NAMELIST /LIST1/ A, B, C /LIST2/ X, Y, Z
```

It is possible to continue a list within the same scoping unit by repeating the namelist name on more than one statement. Thus,

```
NAMELIST /LIST/ A, B, C
NAMELIST /LIST/ D, E, F
```

has the same effect as a single statement containing all the variable names in the same order. A namelist group object may belong to more than one namelist group.

If any variable name in *variable-name-list* appears in a type declaration statement in the same scoping unit, that type declaration statement must either appear before the NAMELIST statement, or confirm the implicit typing rule in force in the scoping unit for the initial letter of the variable. Also, if any variable has the PUBLIC attribute, no variable in the list may have the PRIVATE attribute.

## 7.16  Summary

In this chapter all of the specification statements of Fortran 90 have been described. The following concepts have been introduced: implicit typing and its attendant dangers, named constants, constant expressions, data initialization, control of the accessibility of entities in modules, saving data between procedure calls, selective access of entities in a module, renaming entities accessed from a module, specification expressions that may be used when specifying data objects and function results, and the formation of variables into namelist groups. In addition, we have explained alternative ways of specifying attributes.

The features described here that are new to Fortran are IMPLICIT NONE; initialization and specification expressions; a much extended type declaration statement; DATA statement extended to include derived types, subobjects, and binary, octal, and hexadecimal constants; new attributes and statements: PUBLIC, PRIVATE, POINTER, ALLOCATABLE, TARGET, INTENT, and OPTIONAL; and the USE and NAMELIST statements.

## Exercises

**1.** Write suitable type statements for the following quantities:

i) an array to hold the number of counts in each of the 100 bins of a histogram numbered from 1 to 100;

ii) an array to hold the temperature to two significant decimal places at points, on a sheet of iron, equally spaced at 1cm intervals on a rectangular grid 20cm square, with points in each corner, (the melting point of iron is $1530^{0}C$);

iii) an array to describe the state of 20 on/off switches;

iv) an array to contain the information destined for a printed page of 44 lines each of 70 letters or digits.

**2.** Explain the difference between the following pair of declarations

```
REAL :: I = 3.1
```

and

```
IMPLICIT REAL (I)
PARAMETER (I = 3.1)
```

What is the value of I in each case?

**3.** Write type declaration statements which initialize:

i) all the elements of an integer array of length 100 to the value zero.

ii) all the odd elements of the same array to 0 and the even elements to 1.

iii) the elements of a real 10x10 square array to 1.0 .

iv) a character string to the digits 0 to 9.

**4.** In the following module, identify all the scoping units and list the mappings for implicit typing for all the letters in all of them:

```
MODULE MOD
   IMPLICIT CHARACTER(10, 2) (A-B)
   :
CONTAINS
   SUBROUTINE OUTER
      IMPLICIT NONE
      :
   CONTAINS
     SUBROUTINE INNER(FUN)
        IMPLICIT COMPLEX (Z)
        INTERFACE
           FUNCTION FUN(X)
              IMPLICIT REAL (F, X)
              :
           END FUNCTION FUN
        END INTERFACE
     END SUBROUTINE INNER
   END SUBROUTINE OUTER
END MODULE MOD
```

**5.**

i) Write a type declaration statement that declares and initializes a variable of derived type PERSON (Section 2.9).

ii) Either

    a) write a type declaration statement that declares and initializes a variable of type ENTRY (Section 2.13), or

    b) write a type declaration statement for such a variable and a DATA statement to initialize its nonpointer components.

**6.** Which of the following are initialization expressions:

i) KIND(X), for X of type REAL

  ii) SELECTED_REAL_KIND(6, 20)

 iii) 1.7**2

 iv) 1.7**2.0

  v) (1.7, 2.3)**(–2)

 vi) (/ (7*I, I=1, 10) /)

 vii) PERSON('REID', 25*2., 22**2)

viii) ENTRY(1.7, 1, NULL)

# 8. INTRINSIC PROCEDURES

## 8.1 Introduction

In a language that has a clear orientation towards scientific applications there is an obvious requirement for the most frequently required mathematical functions to be provided as part of the language itself, rather than expecting each user to code them afresh. When provided with the compiler, they are normally coded to be very efficient and will have been well tested over the complete range of values that they accept. It is difficult to compete with the high standard of code provided by the vendors.

The efficiency of the intrinsic procedures when handling arrays on vector or parallel computers is likely to be particularly marked because a single call may cause a large number of individual operations to be performed, during the execution of which advantage may be taken of the specific nature of the hardware.

Another feature of a substantial number of the intrinsic procedures is that they extend the power of the language by providing access to facilities that are not otherwise available in the language. Examples are inquiry functions for the presence of an optional argument, the parts of a floating-point number, and the length of a character string.

There are over a hundred intrinsic procedures in all, a particularly rich set. They fall into distinct groups, which we describe in turn. A list in alphabetical order, with one-line descriptions, is given in Appendix A. Some processors may offer additional intrinsic procedures. Note that a program containing references to such procedures does not conform to the standard, and is portable only to other processors that provide those same procedures.

## 8.1.1 Keyword calls

The procedures may be called with keyword actual arguments, using the dummy argument names as keywords. This facility is not very useful for those with a single non-optional argument, but is useful for those with several optional arguments. For example

```
CALL DATE_AND_TIME (DATE=D)
```

returns the date in the scalar character variable D.

Some procedures have all their arguments optional although every call can be expected to have at least one argument present; for these procedures we have taken a little 'poetic licence' with the bracket notation since there must not be a leading comma if the first argument is omitted.

## 8.1.2  Categories of intrinsic procedures

There are four categories of intrinsic procedures:

i) *Elemental procedures* are specified for scalar arguments, but may also be applied to conforming array arguments.  In the case of a function, each element of the result, if any, is as would have been obtained by applying the function to corresponding elements of each of the array arguments.  In the case of a subroutine,[13] each argument of intent OUT or INOUT must be an array, and each element is as would have resulted from applying the subroutine to corresponding elements of each of the array arguments.

ii) *Inquiry functions* return properties of their principal arguments that do not depend on their values; indeed their values may be undefined.

iii) *Transformational functions* are functions that are neither elemental nor inquiry; they usually have array arguments and an array result whose elements depend on many of the elements of the arguments.

iv) *Nonelemental subroutines.*

## 8.1.3  INTRINSIC statement

A name may be specified to be that of an intrinsic procedure in an INTRINSIC statement, which has the general form

INTRINSIC *intrinsic-name-list*

where *intrinsic-name-list* is a list of intrinsic procedure names.  A name must not appear more than once in the INTRINSIC statements of a scoping unit and must not appear in an EXTERNAL statement there.  We believe that it is good programming practice to include such a statement in every scoping unit that contains references to intrinsic procedures, because this makes the use clear to the reader.  We particularly recommend it when referencing intrinsic procedures that are not defined by the standard, for then a clear diagnostic message should be produced if the program is ported to a processor that does not support the extra intrinsic procedures.

---

[13] In fact, Fortran 90 has only one elemental subroutine.

## 8.2  Inquiry functions for any type

The following are inquiry functions whose arguments may be of any type:

**ASSOCIATED (POINTER [,TARGET]),** when TARGET is absent, returns the value .TRUE. if the pointer POINTER is associated with a target and .FALSE. otherwise.  The pointer association status of POINTER must not be undefined.  If TARGET is present, the value is .TRUE. if POINTER is associated with TARGET, and .FALSE. otherwise. TARGET may itself be a pointer, in which case its target is compared with the target of POINTER; the pointer association status of TARGET must not be undefined and if either POINTER or TARGET is disassociated, the result is .FALSE. .

**PRESENT (A)** may be called in a subprogram that has an optional dummy argument A.  It returns the logical value .TRUE. if the corresponding actual argument is present in the current call of the subprogram, and .FALSE. otherwise.  If an absent dummy argument is used as an actual argument in a call of another subprogram, it is regarded as also absent in the called subprogram.

The following is an inquiry function whose argument may be of any intrinsic type:

**KIND (X)** has type default integer and value equal to the kind type parameter value of X.

## 8.3  Elemental numeric functions

There are seventeen elemental functions for performing simple numerical tasks, many of which perform type conversions for some or all types of arguments.

### 8.3.1  Elemental functions that may convert

If KIND is present in the following elemental functions, it must be a scalar integer initialization expression.

**ABS (A)** returns the absolute value of an argument of type integer, real, or complex.  The result is of type integer if A is of type integer and otherwise it is of type real.  It has the same kind type parameter as A.

**AIMAG (Z)** returns the imaginary part of the complex value Z.  The type is real and the kind type parameter is that of Z.

**AINT (A [,KIND])** truncates a real value A towards zero to produce a real that is a whole number. The value of the kind type parameter is the value of the argument KIND if it is present, or that of the default real otherwise.

**ANINT (A [,KIND])** returns a real whose value is the nearest whole number to the real value A. The value of the kind type parameter is the value of the argument KIND, if it is present, or that of the default real otherwise.

**CEILING (A)** returns the least default integer greater than or equal to its real argument.

**CMPLX (X [,Y] [,KIND])** converts X or (X, Y) to complex type with the value of the kind type parameter being the value of the argument KIND if it is present or that of the default complex otherwise. If Y is absent, X may be of type integer, real, or complex. If Y is present, it must be of type integer or real and X must be of type integer or real.

**FLOOR (A)** returns the greatest default integer less than or equal to its real argument.

**INT (A [,KIND])** converts to integer type with the value of the kind type parameter being the value of the argument KIND, if it is present, or that of the default integer otherwise. A may be

- integer, in which case INT(A)=A,

- real, in which case the value is truncated towards zero, or

- complex, in which case the real part is truncated towards zero.

**NINT (A [,KIND])** returns the integer value that is nearest to the real A. If KIND is present, the value of the kind type parameter of the result is the value of KIND, otherwise it is that of the default integer type.

**REAL (A [,KIND])** converts to real type with the value of the kind type parameter being that of KIND if it is present. If KIND is absent, the kind type parameter is that of default real when A is of type integer or real, and is that of A when A is type complex. A may be of type integer, real, or complex. If it is complex, the imaginary part is ignored.

## 8.3.2  Elemental functions that do not convert

The following are elemental functions whose result is of type and kind type parameter that are those of the first or only argument. For those having more than one argument, all arguments must have the same type and kind type parameter.

**CONJG (Z)** returns the conjugate of the complex value Z.

**DIM (X, Y)** returns max(X–Y, 0.) for arguments that are both integer or both real.

**MAX (A1, A2 [,A3,...])** returns the maximum of two or more integer or real values.

**MIN (A1, A2 [,A3,...])** returns the minimum of two or more integer or real values.

**MOD (A, P)** returns the remainder of A modulo P, that is A–INT(A/P) * P. If P=0, the result is processor dependent. A and P must be both integer or both real.

**MODULO (A, P)** returns A modulo P for A and P both integer or both be real, that is A–FLOOR(A/P)*P in the real case, and A–FLOOR(REAL(A)/REAL(P))*P in the integer case. If P=0, the result is processor dependent.

**SIGN (A, B)** returns the absolute value of A times the sign of B. A and B must be both integer or both real. If B=0, its sign is taken as positive.

## 8.4  Elemental mathematical functions

The following are elemental functions that evaluate elementary mathematical functions. The type and kind type parameter of the result are those of the first argument, which is usually the only argument.

**ACOS (X)** returns the arc cosine (inverse cosine) function value for real values X such that $|X| \leq 1$, expressed in radians in the range $0 \leq ACOS(X) \leq \pi$.

**ASIN (X)** returns the arc sine (inverse sine) function value for real values X such that $|X| \leq 1$, expressed in radians in the range $-\pi/2 \leq ASIN(X) \leq \pi/2$.

**ATAN (X)** returns the arc tangent (inverse tangent) function value for real X, expressed in radians in the range $-\pi/2 \leq ATAN(X) \leq \pi/2$.

**ATAN2 (Y, X)** returns the arc tangent (inverse tangent) function value for pairs of reals, X and Y, of the same type and type parameter. The result is the principal value of the argument of the complex number (X,Y), expressed in radians in the range $-\pi <$ ATAN2(Y,X) $\leq \pi$. The values of X and Y must not both be zero.

**COS (X)** returns the cosine function value for an argument of type real or complex that is treated as a value in radians.

**COSH (X)** returns the hyperbolic cosine function value for a real argument X.

**EXP (X)** returns the exponential function value for a real or complex argument X.

**LOG (X)** returns the natural logarithm function for a real or complex argument X. In the real case, X must be positive. In the complex case, X must not be zero, and the imaginary part $w$ of the result lies in the range $-\pi < w \leq \pi$.

**LOG10 (X)** returns the common (base 10) logarithm of a real argument whose value must be positive.

**SIN (X)** returns the sine function value for a real or complex argument that is treated as a value in radians.

**SINH (X)** returns the hyperbolic sine function value for a real argument.

**SQRT (X)** returns the square root function value for a real or complex argument X. If X is real, its value must be not be negative. In the complex case, the real part of the result is not negative, and when it is zero the imaginary part of the result is not negative.

**TAN (X)** returns the tangent function value for a real argument that is treated as a value in radians.

**TANH (X)** returns the hyperbolic tangent function value for a real argument.

## 8.5  Elemental character and logical functions

### 8.5.1  Character-integer conversions

The following are elemental functions for conversions from a single character to an integer, and vice-versa.

**ACHAR (I)** is of type default character with length one and returns the character in the position in the ASCII collating sequence that is specified by the integer I. I must be in the range $0 \leq I \leq 127$, otherwise the result is processor dependent.

**CHAR (I [,KIND])** is of type character and length one, with a kind type parameter value that of the value of KIND if present, or default otherwise.   It returns the character in position I in the processor collating sequence associated with the relevant kind parameter.  I must be in the range $0 \le I \le n-1$, where $n$ is the number of characters in the processor's collating sequence.  If KIND is present, it must be a scalar integer initialization expression.

**IACHAR (C)** is of type default integer and returns the position in the ASCII collating sequence of the default character C.   If C is not in the sequence, the result is processor dependent.

**ICHAR (C)** is of type default integer and returns the position of the character C in the processor collating sequence associated with the kind parameter of C.

## 8.5.2 Lexical comparison functions

The following elemental functions accept default character strings as arguments, make a lexical comparison based on the ASCII collating sequence, and return a default logical result.  If the strings have different lengths, the shorter one is padded on the right with blanks.

**LGE (STRING_A, STRING_B)** returns the value .TRUE. if STRING_A follows STRING_B in the ASCII collating sequence or is equal to it, and the value .FALSE. otherwise.

**LGT (STRING_A, STRING_B)** returns the value .TRUE. if STRING_A follows STRING_B in the ASCII collating sequence, and the value .FALSE. otherwise.

**LLE (STRING_A, STRING_B)** returns the value .TRUE. if STRING_A precedes STRING_B in the ASCII collating sequence or is equal to it, and the value .FALSE. otherwise.

**LLT (STRING_A, STRING_B)** returns the value .TRUE. if STRING_A precedes STRING_B in the ASCII collating sequence, and .FALSE. otherwise.

## 8.5.3 String-handling elemental functions

The following are elemental functions that manipulate strings.  The arguments STRING, SUBSTRING, and SET are always of type character, and where two are present have the same kind type parameter.  The kind type parameter value of the result is that of STRING.

**ADJUSTL (STRING)** adjusts left to return a string of the same length by removing all leading blanks and inserting the same number of trailing blanks.

**ADJUSTR (STRING)** adjusts right to return a string of the same length by removing all trailing blanks and inserting the same number of leading blanks.

**INDEX (STRING, SUBSTRING [,BACK])** has type default integer and returns the starting position of SUBSTRING as a substring of STRING, or zero if it does not occur as a substring. If BACK is absent or present with value false, the starting position of the first such substring is returned; the value 1 is returned if SUBSTRING has zero length. If BACK is present with value true, the starting position of the last such substring is returned; the value LEN(STRING)+1 is returned if SUBSTRING has zero length.

**LENTRIM (STRING)** returns a default integer whose value is the length of STRING without trailing blank characters.

**SCAN (STRING, SET [,BACK])** returns a default integer whose value is the position of a character of STRING that is in SET, or zero if there is no such character. If the logical BACK is absent or present with value false, the position of the leftmost such character is returned. If BACK is present with value true, the position of the rightmost such character is returned.

**VERIFY (STRING, SET [,BACK])** returns the default integer value 0 if each character in STRING appears in SET, or the position of a character of STRING that is not in SET. If the logical BACK is absent or present with value false, the position of the left-most such character is returned. If BACK is present with value true, the position of the rightmost such character is returned.

## 8.5.4 Logical conversion

The following elemental function converts from a logical value with one kind type parameter to another.

**LOGICAL (L [,KIND] )** returns a logical value equal to the value of the logical L. The value of the kind type parameter of the result is the value of KIND if it is present or that of default logical otherwise. If KIND is present, it must be a scalar integer initialization expression.

## 8.6  Non-elemental string-handling functions

### 8.6.1  String-handling inquiry function

**LEN (STRING)** is an inquiry function that returns a scalar default integer holding the number of characters in STRING if it is scalar or in an element of STRING if it is array valued.  The value of STRING need not be defined.

### 8.6.2  String-handling transformational function

There are two functions that cannot be elemental because the length type parameter of the result depends on the value of an argument.

**REPEAT (STRING, NCOPIES)** forms the string consisting of the concatenation of NCOPIES copies of STRING, where NCOPIES is of type integer and its value must not be negative.  Both arguments must be scalar.

**TRIM (STRING)** returns STRING with all trailing blanks removed.  STRING must be scalar.

## 8.7  Numeric inquiry and manipulation functions

### 8.7.1  Models for integer and real data

The numeric inquiry and manipulation functions are defined in terms of a model set of integers and a model set of reals for each kind of integer and real data type implemented.  For each kind of integer, it is the set

$$i = s \times \sum_{k=1}^{q} w_k \times r^{k-1}$$

where $s$ is $\pm 1$, $q$ is a positive integer, $r$ is an integer exceeding one (usually 2), and each $w_k$ is an integer in the range $0 \le w_k < r$.  For each kind of real, it is the set

$$x = 0$$

and

$$x = s \times b^e \times \sum_{k=1}^{p} f_k \times b^{-k}$$

where $s$ is $\pm 1$, $p$ and $b$ are integers exceeding one, $e$ is an integer in a range $e_{min} \le e \le e_{max}$, and each $f_k$ is an integer in the range $0 \le f_k < b$ except that $f_1$ is also nonzero.

Values of the parameters in these models are chosen for the processor so as best to fit the hardware with the proviso that all model numbers are representable. Note that it is quite likely that there are some machine numbers that lie outside the model. For example, many computers represent the integer $-r^q$ and the IEEE Standard for Binary Floating-point Arithmetic contains reals with $f_1 = 0$ (called denormalized numbers) and register numbers with increased precision and range.

## 8.7.2 Numeric inquiry functions

There are nine inquiry functions that return values from the models associated with their arguments. Each has a single argument that may be scalar or array-valued and each returns a scalar result. The value of the argument need not be defined.

**DIGITS (X),** for real or integer X, returns the default integer whose value is the number of significant digits in the model that includes X, that is $p$ or $q$.

**EPSILON (X),** for real X, returns a real result with the same type parameter as X that is almost negligible compared with the value one in the model that includes X, that is $b^{1-p}$.

**HUGE (X),** for real or integer X, returns the largest value in the model that includes X. It has the type and type parameter of X. The value is

$$(1-b^{-p})b^{e_{max}}$$

or

$$r^q-1$$

**MAXEXPONENT (X),** for real X, returns the default integer $e_{max}$, the maximum exponent in the model that includes X.

**MINEXPONENT (X),** for real X, returns the default integer $e_{min}$, the minimum exponent in the model that includes X.

**PRECISION (X),** for real or complex X, returns a default integer holding the equivalent decimal precision in the model representing real numbers with the same type parameter value as X. The value is INT($(p-1)$*LOG10($b$))+$k$, where $k$ is 1 if $b$ is an integral power of 10 and 0 otherwise.

**RADIX (X),** for real or integer X, returns the default integer that is the base in the model that includes X, that is $b$ or $r$.

**RANGE (X),** for integer, real, or complex X, returns a default integer holding the equivalent decimal exponent range in the models representing integer or real numbers with the same type parameter value as X. The value is INT(LOG10($huge$)) for integers and

$$INT(MIN(LOG10(huge), -LOG10(tiny)))$$

for reals, where $huge$ and $tiny$ are the largest and smallest numbers in the models.

**TINY (X),** for real X, returns the smallest positive number

$$b^{e_{min}-1}$$

in the model that includes X. It has the type and type parameter of X.

## 8.7.3 Elemental functions to manipulate reals

There are seven elemental functions whose first or only argument is of type real and that return values related to the components of the model values associated with the actual value of the argument.

**EXPONENT (X)** returns the default integer whose value is the exponent part $e$ of X when represented as a model number. If X=0, the result has value zero.

**FRACTION (X)** returns a real with the same type parameter as X whose value is the fractional part of X when represented as a model number, that is $X\, b^{-e}$.

**NEAREST (X, S)** returns a real with the same type parameter as X whose value is the nearest different machine number in the direction given by the sign of the real S. The value of S must not be zero.

**RRSPACING (X)** returns a real with the same type parameter as X whose value is the reciprocal of the relative spacing of model numbers near X, that is $|\,X\, b^{-e}|\, b^p$.

**SCALE (X, I)** returns a real with the same type parameter as X, whose value is $X\, b^I$, where $b$ is the base in the model for X, and I is of type integer.

**SET_EXPONENT (X, I)** returns a real with the same type parameter as X, whose fractional part is the fractional part of the model representation of X and whose exponent part is I, that is $X\, b^{I-e}$.

SPACING (X) returns a real with the same type parameter as X whose value is the absolute spacing of model numbers near X, that is $b^{e-p}$.

## 8.7.4 Transformational functions for kind values

There are two functions that return the least kind type parameter value that will meet a given numeric requirement. They have scalar arguments and results, so are classified as transformational.

SELECTED_INT_KIND (R) returns the default integer scalar that is the kind type parameter value for an integer data type able to represent all integer values $n$ in the range $-10^R < n < 10^R$, where R is a scalar integer. If more than one is available, a kind with least decimal exponent range is chosen (and least kind value if several have least decimal exponent range). If no corresponding kind is available, the result is $-1$.

SELECTED_REAL_KIND ([P][, R]) returns the default integer scalar that is the kind type parameter value for a real data type with decimal precision (as returned by the function PRECISION) at least P, and decimal exponent range (as returned by the function RANGE) at least R. If more than one is available, a kind with the least decimal precision is chosen (and least kind value if several have least decimal precision). Both P and R are scalar integers. At least one of them must be present. If no corresponding kind value is available, the result is $-1$ if the requested precision is unavailable, $-2$ if the requested exponent range is unavailable, and $-3$ if both are unavailable.

## 8.8 Bit manipulation procedures

There are eleven procedures for manipulating bits held within integers. They are based on those in the US Military Standard MIL-STD 1753. They differ only in that here they are elemental, where appropriate, whereas the original procedures accepted only scalar arguments.

These intrinsics are based on a model in which an integer holds $s$ bits $w_k$, $k = 0, 1 ,..., s-1$, in a sequence from right to left, based on the nonnegative value

$$\sum_{k=0}^{s-1} w_k \times 2^k$$

This model is valid only in the context of these intrinsics. It is identical to the model for integers in Section 8.7.1 when $r$ is an integral power of 2 and

$w_{s-1} = 0$, but when $w_{s-1} = 1$ the models do not correspond, and the value expressed as an integer may vary from processor to processor.

## 8.8.1 Inquiry function

**BIT_SIZE (I)** returns the number of bits in the model for bits within an integer of the same type parameter as I.    The result is a scalar integer having the same type parameter as I.

## 8.8.2 Elemental functions

**BTEST (I, POS)** returns the default logical value .TRUE. if bit POS of the integer I has value 1 and .FALSE. otherwise.    POS must be an integer with value in the range $0 \le POS < BIT\_SIZE(I)$.

**IAND (I, J)** returns the logical AND of all the bits in I and corresponding bits in J, according to the truth table

```
I             1   1   0   0
J             1   0   1   0
IAND(I, J)    1   0   0   0
```

I and J must have the same type parameter value, which is the type parameter value of the result.

**IBCLR (I, POS)** returns an integer, with the same type parameter as I, and value equal to that of I except that bit POS is changed to 0.  POS must be an integer with value in the range $0 \le POS < BIT\_SIZE(I)$.

**IBITS (I, POS, LEN)** returns an integer, with the same type parameter as I, and value equal to the LEN bits of I starting at bit POS right adjusted and all other bits zero.  POS and LEN must be integers with nonnegative values such that $POS+LEN \le BIT\_SIZE(I)$.

**IBSET (I, POS)** returns an integer, with the same type parameter as I, and value equal to that of I except that bit POS is changed to 1.  POS must be an integer with value in the range $0 \le POS < BIT\_SIZE(I)$.

**IEOR (I, J)** returns the logical exclusive OR of all the bits in I and corresponding bits in J, according to the truth table

```
I             1   1   0   0
J             1   0   1   0
IEOR(I, J)    0   1   1   0
```

I and J must have the same type parameter value, which is the type parameter value of the result.

**IOR (I, J)** returns the logical inclusive OR of all the bits in I and corresponding bits in J, according to the truth table

```
I              1  1  0  0
J              1  0  1  0
IOR(I, J)      1  1  1  0
```

I and J must have the same type parameter value, which is the type parameter value of the result.

**ISHFT (I, SHIFT)** returns an integer, with the same type parameter as I, and value equal to that of I except that the bits are shifted SHIFT places to the left (–SHIFT places to the right if SHIFT is negative). Zeros are shifted in from the other end. SHIFT must be an integer with value satisfying the inequality | SHIFT | ≤ BIT_SIZE(I).

**ISHFTC (I, SHIFT [, SIZE])** returns an integer, with the same type parameter as I, and value equal to that of I except that the SIZE rightmost bits (or all the bits if SIZE is absent) are shifted circularly SHIFT places to the left (–SHIFT places to the right if SHIFT is negative). SHIFT must be an integer with absolute value not exceeding the value of SIZE (or BIT_SIZE(I) if SIZE is absent).

**NOT (I)** returns the logical complement of all the bits in I, according to the truth table

```
I          0  1
NOT(I)     1  0
```

## 8.8.3 Elemental subroutine

**CALL MVBITS (FROM, FROMPOS, LEN, TO, TOPOS)** copies the sequence of bits in FROM that starts at position FROMPOS and has length LEN to TO, starting at position TOPOS. The other bits of TO are not altered. FROM, FROMPOS, LEN, and TOPOS are all integers with intent IN, and they must have values that satisfy the inequalities: LEN ≥ 0, FROMPOS ≥ 0, FROMPOS+LEN ≤ BIT_SIZE(FROM), TOPOS ≥ 0, and TOPOS+LEN ≤ BIT_SIZE(TO). TO is an integer with intent INOUT; it must have the same kind type parameter as FROM. The same variable may be specified for FROM and TO.

## 8.9 Transfer function

The transfer function allows data of one type to be transferred to another without the physical representation being altered. This would be useful, for example, in writing a data storage and retrieval system. The system itself could be written for one type, default integer say, and other types handled by transfers to and from that type.

**TRANSFER (SOURCE, MOLD [,SIZE])** returns a result of type and type parameters those of MOLD. When SIZE is absent, the result is scalar if MOLD is scalar, and it is of rank one and size just sufficient to hold all of SOURCE if MOLD is array-valued. When SIZE is present, the result is of rank one and size SIZE. If the physical representation of the result is as long as or longer than that of SOURCE, the result contains SOURCE as its leading part and the rest is undefined; otherwise the result is the leading part of SOURCE.

## 8.10 Vector and matrix multiplication functions

There are two transformational functions that perform vector and matrix multiplications. They each have two arguments that are both of numeric type (integer, real, or complex) or both of logical type. The result is of the same type and type parameter as for the multiply or AND operation between two such scalars. The functions SUM and ANY, used in the definitions, are defined in Section 8.11.1.

**DOT_PRODUCT (VECTOR_A, VECTOR_B)** requires two arguments each of rank one and the same size. It returns SUM(VECTOR_A * VECTOR_B) if VECTOR_A is of type integer or type real, it returns SUM(CONJG(VECTOR_A) * VECTOR_B) if VECTOR_A is of type complex, and it returns ANY(VECTOR_A .AND. VECTOR_B) if VECTOR_A is of type logical.

**MATMUL (MATRIX_A, MATRIX_B)** performs matrix multiplication. For numeric arguments, three cases are possible:

   i) MATRIX_A has shape $(n, m)$ and MATRIX_B has shape $(m, k)$. The result has shape $(n, k)$ and element $(i, j)$ has value SUM(MATRIX_A$(i, :)$ * MATRIX_B$(:, j)$).

   ii) MATRIX_A has shape $(m)$ and MATRIX_B has shape $(m, k)$. The result has shape $(k)$ and element $(j)$ has value SUM(MATRIX_A * MATRIX_B$(:, j)$).

iii) MATRIX_A has shape ($n,m$) and MATRIX_B has shape ($m$). The result has shape ($n$) and element ($i$) has value SUM(MATRIX_A($i$, :) * MATRIX_B).

For logical arguments, the shapes are as for numeric arguments and the values are determined by replacing 'SUM' and '*' in the above expressions by 'ANY' and '.AND.'.

## 8.11  Transformational functions that reduce arrays

There are seven transformational functions that perform operations on arrays such as summing their elements.

### 8.11.1  Single argument case

In their simplest form, these functions have a single array argument and return a scalar result. All except COUNT have a result of the same type and type parameter as the argument.

**ALL (MASK)** returns the value true if all elements of the logical array MASK are true or MASK has size zero, and otherwise returns the value false.

**ANY (MASK)** returns the value true if any of the elements of the logical array MASK is true, and returns the value false if no elements are true or if MASK has size zero.

**COUNT (MASK)** returns the default integer value that is the number of elements of the logical array MASK that have the value true.

**MAXVAL (ARRAY)** returns the maximum value of an element of an integer or real array. If ARRAY has size zero, it returns the negative value of largest magnitude supported by the processor.

**MINVAL (ARRAY)** returns the minimum value of an element of an integer or real array. If ARRAY has size zero, it returns the largest positive value supported by the processor.

**PRODUCT (ARRAY)** returns the product of the elements of an integer, real, or complex array. It returns the value one if ARRAY has size zero.

**SUM (ARRAY)** returns the sum of the elements of an integer, real, or complex array. It returns the value zero if ARRAY has size zero.

## 8.11.2  Optional argument DIM

All these functions have an optional second argument DIM that is a scalar integer.  If this is present, the operation is applied to all rank-one sections that span right through dimension DIM to produce an array of rank reduced by one and extents equal to the extents in the other dimensions.  For example, if A is a real array of shape (4,5,6), SUM(A,DIM=2) is a real array of shape (4,6) and element $(i, j)$ has value SUM(A($i$, :, $j$)).

## 8.11.3  Optional argument MASK

The functions MAXVAL, MINVAL, PRODUCT, and SUM have a third optional argument, a logical array MASK.  If this is present, it must have the same shape as the first argument and the operation is applied to the elements corresponding    to    true    elements    of    MASK;    for    example, SUM(A, MASK = A>0) sums the positive elements of the array A.  MASK affects only the value of the function and does not affect the evaluation of arguments that are array expressions.

## 8.12  Array inquiry functions

There are five functions for inquiries about the bounds, shape, size and allocation status of an array of any type.  Because the result depends only the array properties, the value of the array need not be defined.

## 8.12.1  Allocation status

**ALLOCATED (ARRAY)** returns, when the allocatable array ARRAY is currently allocated, the value .TRUE.; otherwise it returns the value .FALSE. .   If the allocation status of ARRAY is undefined, the result is undefined.

## 8.12.2  Bounds, shape, and size

The following functions enquire about the bounds of an array.  In the case of an allocatable array, it must be allocated; and in the case of a pointer array, it must be associated with a target.  An array section or an array expression is taken to have lower bounds 1 and upper bounds equal to the extents.

**LBOUND (ARRAY [,DIM])**, when DIM is absent, returns a rank-one default integer array holding the lower bounds.  When DIM is present, it must be a scalar integer and the result is a scalar default integer holding the lower bound in dimension DIM.

**SHAPE (SOURCE)** returns a rank-one default integer array holding the shape of the array or scalar SOURCE. In the case of a scalar, the result has size zero.

**SIZE (ARRAY [,DIM])** returns a scalar default integer that is the size of the array ARRAY or extent along dimension DIM if the scalar integer DIM is present.

**UBOUND (ARRAY [,DIM])** is similar to LBOUND except that it returns upper bounds.

## 8.13  Array construction and manipulation functions

There are eight functions that construct or manipulate arrays of any type.

### 8.13.1  Merge elemental function

**MERGE (TSOURCE, FSOURCE, MASK)**    is    an    elemental    function. TSOURCE may have any type and FSOURCE must have the same type and type parameters. MASK must be of type logical. The result is TSOURCE if MASK is true and FSOURCE otherwise.

The principal application of MERGE is when the three arguments are arrays having the same shape, in which case TSOURCE and FSOURCE are merged under the control of MASK. Note, however, that TSOURCE or FSOURCE may be scalar in which case the elemental rules effectively broadcast it to an array of the correct shape.

### 8.13.2  Packing and unpacking arrays

The transformational function PACK packs into a rank-one array those elements of an array that are selected by a logical array of conforming shape, and the transformational function UNPACK performs the reverse operation. The elements are taken in array element order.

**PACK (ARRAY, MASK [,VECTOR]),** when VECTOR is absent, returns a rank-one array containing the elements of ARRAY corresponding to true elements of MASK in array element order; MASK may be scalar with value true, in which case all elements are selected. If VECTOR is present, it must be a rank-one array of the same type and type parameters as ARRAY and size at least equal to the number $t$ of selected elements; the result has size equal to the size $n$ of VECTOR; if $t < n$, elements $i$ of the result for $i > t$ are the corresponding elements of VECTOR.

**UNPACK (VECTOR, MASK, FIELD)** returns an array of the type and type parameters of VECTOR and shape of MASK. MASK must be a logical array and VECTOR must be a rank-one array of size at least the number of true elements of MASK. FIELD must be of the same type and type parameters as VECTOR and must either be scalar or be of the same shape as MASK. The element of the result corresponding to the $i$th true element of MASK, in array-element order, is the $i$th element of VECTOR; all others are equal to the corresponding elements of FIELD if it is an array or to FIELD if it is a scalar.

## 8.13.3  Reshaping an array

The transformational function RESHAPE allows the shape of an array to be changed, with possible permutation of the subscripts.

**RESHAPE (SOURCE, SHAPE [,PAD] [,ORDER])** returns an array with shape given by the rank-one integer array SHAPE, and type and type parameters those of the array SOURCE. The size of SHAPE must be constant, and its elements must not be negative. If PAD is present it must be an array of the same type and type parameters as SOURCE. If PAD is absent or of size zero, the size of the result must not exceed the size of SOURCE. If ORDER is absent, the elements of the result, in array element order, are the elements of SOURCE in array element order followed by copies of PAD in array-element order. If ORDER is present, it must be an integer array of the same shape as SHAPE, and the elements $R(s_1,..., s_n)$ of the result, taken in subscript order for the array having elements $R(s_{ORDER(1)},..., s_{ORDER(n)})$, are those of SOURCE in array element order followed by copies of PAD in array-element order. The value of ORDER must be a permutation of $(1,2,...,n)$.

## 8.13.4  Transformational function for replication

**SPREAD (SOURCE, DIM, NCOPIES)** returns an array of type and type parameters those of SOURCE and of rank increased by one. SOURCE may be scalar or array-valued. DIM and NCOPIES are integer scalars. The result contains MAX(NCOPIES, 0) copies of SOURCE, and element $(r_1,...,r_{n+1})$ of the result is SOURCE$(s_1,...,s_n)$ where $(s_1,...,s_n)$ is $(r_1,...,r_{n+1})$ with subscript DIM omitted (or SOURCE itself if it is scalar).

### 8.13.5  Array shifting functions

CSHIFT (ARRAY, SHIFT [,DIM]) returns an array of the same type, type parameters, and shape as ARRAY. DIM is an integer scalar. If DIM is omitted, it is as if it were present with the value 1. SHIFT is of type integer and must be scalar if ARRAY is of rank one. If SHIFT is scalar, the result is obtained by shifting every rank-one section that extends across dimension DIM circularly SHIFT times. The direction of the shift depends on the sign of SHIFT and may be determined from the case with SHIFT=1 and ARRAY of rank one and size $m$, when element $i$ of the result is ARRAY$(i + 1)$, $i$=1,2,...,$m$−1 and element $m$ is ARRAY(1). If SHIFT is an array, it must have shape that of ARRAY with dimension DIM omitted, and it supplies a separate value for each shift.

EOSHIFT (ARRAY, SHIFT [,BOUNDARY] [,DIM]) is identical to CSHIFT except that an end-off shift is performed and boundary values are inserted into the gaps so created. BOUNDARY may be omitted when ARRAY has intrinsic types, in which case the value zero is inserted for the integer, real, and complex cases; false in the logical case; and blanks in the character case. If BOUNDARY is present, it must have the same type and type parameters as ARRAY; it may be scalar and supply all needed values or it may be an array whose shape is that of ARRAY with dimension DIM omitted and supply a separate value for each shift.

### 8.13.6  Matrix transpose

The TRANSPOSE function performs a matrix transpose for any array of rank two.

TRANSPOSE (MATRIX) returns an array of the same type and type parameters as the rank-two array MATRIX. Element $(i, j)$ of the result is MATRIX$(j, i)$.

### 8.14  Transformational functions for geometric location

There are two transformational functions that find the locations of the maximum and minimum values of an integer or real array.

MAXLOC (ARRAY [,MASK]) returns a rank-one default integer array of size equal to the rank of ARRAY. Its value is the sequence of subscripts of an element of maximum value (among those corresponding to true values of the conforming logical array MASK if it is present), as though all the declared lower bounds of ARRAY were 1. If there is more than one such element, the first in array element order is taken.

**MINLOC (ARRAY [,MASK])** returns a rank-one default integer array of size equal to the rank of ARRAY. Its value is the sequence of subscripts of an element of minimum value (among those corresponding to true values of the conforming logical array MASK if it is present), as though all the declared lower bounds of ARRAY were 1. If there is more than one such element, the first in array element order is taken.

## 8.15  Nonelemental intrinsic subroutines

There are four nonelemental intrinsic subroutines, which were chosen to be subroutines rather than functions because of the need to return information through the arguments.

## 8.15.1  Real-time clock

There are two subroutines that return information from the real-time clock, the first based on the ISO standard IS 8601 (Representation of dates and times). It is assumed that there is a basic system clock that is incremented by one for each clock count until a maximum COUNT_MAX is reached and on the next count is set to zero. Default values are returned on systems without a clock. All the arguments have intent OUT.

**CALL DATE_AND_TIME ([DATE] [,TIME] [,ZONE] [,VALUES])** returns the following (with default values blank or –HUGE(0), as appropriate, when there is no clock):

DATE is a scalar default character variable holding the date in the form *ccyymmdd*, corresponding to century, year, month, and day.

TIME is a scalar default character variable holding the time in the form *hhmmss.sss*, corresponding to hours, minutes, seconds, and milliseconds.

ZONE is a scalar default character variable that is set to the difference between local time and Coordinated Universal Time (UTC, also known as Greenwich Mean Time) in the form *Shhmm*, corresponding to sign, hours, and minutes. For example, a processor in New York in winter would return the value –0500.

VALUES is a rank-one default integer array holding the sequence of values: the year, the month of the year, the day of the month, the time difference in minutes with respect to UTC, the hour of the day, the minutes of the hour, the seconds of the minute, and the milliseconds of the second.

**CALL SYSTEM_CLOCK ([COUNT] [,COUNT_RATE] [,COUNT_MAX])**
returns the following:

COUNT is a scalar default integer holding the current value of the system clock, or $-$HUGE(0) if there is no clock.

COUNT_RATE is a scalar default integer holding the number of clock counts per second, or zero if there is no clock.

COUNT_MAX is a scalar default integer holding the maximum value that COUNT may take, or zero if there is no clock.

## 8.15.2  Random numbers

Pseudorandom numbers are generated from a seed that is held as a rank-one array of integers.   The subroutine RANDOM_NUMBER returns the pseudorandom numbers and the subroutine RANDOM_SEED allows an inquiry to be made about the size or value of the seed array, and the seed to be reset.

**CALL RANDOM_NUMBER (HARVEST)** returns a pseudorandom number from the uniform distribution over the range $0 \le x < 1$ or an array of such numbers.   HARVEST has intent OUT, may be scalar or array-valued, and must be of type real.

**CALL RANDOM_SEED ([SIZE] [,PUT] [,GET])** has the following arguments:

SIZE has intent OUT and is a scalar default integer that the processor sets to the size N of the seed array.

PUT has intent IN and is a default integer array of rank one and size N that is used by the processor to reset the seed.

GET has intent OUT and is a default integer array of rank one and size N that the processor sets to the current value of the seed.

## 8.16 Summary

In this chapter, we introduced the four categories of intrinsic procedures, explained the INTRINSIC statement, and gave detailed descriptions of all the procedures. The procedures of Sections 8.2, 8.6.1, and 8.7 to 8.15 are all new to Fortran.  Within Sections 8.3 to 8.5 the procedures CEILING, FLOOR, MODULO, ACHAR, IACHAR, ADJUSTL, ADJUSTR, LENTRIM, SCAN, VERIFY, and LOGICAL are new; the remaining functions were present in Fortran 77, have been generalized to handle all kind type parameters, have become elemental, and several have been given additional optional arguments. The function LEN has become an inquiry function.

## Exercises

**1.** Write a program to calculate the real or imaginary roots of the quadratic equation $ax^2 + bx + c = 0$ for any values of $a$, $b$, and $c$.  The program should read these three values and print the results.  Use should be made of the appropriate intrinsic functions.

**2.** Repeat Exercise 1 of Chapter 5, avoiding the use of DO constructs.

# 9. DATA TRANSFER

## 9.1 Introduction

Fortran 90 has, in comparison with most other high-level programming languages, a particularly rich set of facilities for input/output (I/O). The 1978 standard brought with it important new features including direct-access files, internal files, execution-time format specification, list-directed input/output, file inquiry, and some new edit descriptors. By contrast, the only significant new features in the latest standard are non-advancing I/O, NAMELIST, and some new edit descriptors. In addition, there are a number of detailed changes to support new facilities in other areas.

Input/output is an area of Fortran 90 into which not all programmers need to delve very deeply. For most small-scale programs it is sufficient to know how to read a few data records containing input variables, and how to transmit to a terminal or printer the results of a calculation. In large-scale data processing, on the other hand, the programs often have to deal with huge streams of data to and from many disc, tape, and cartridge files; in these cases it is essential that great attention be paid to the way in which the I/O is designed and coded, as otherwise both the execution time and the real time spent in the program can suffer dramatically. The term *file* is used for a collection of data on one of these devices and a file is always organized into a sequence of *records*.

This chapter begins by discussing the various forms of formatted I/O, that is I/O which deals with records that do not use the internal number representation of the computer, but rather a character string which can be displayed for visual inspection by the human eye. It is also the form usually needed for transmitting data between different kinds of computers. The so-called *edit descriptors*, which are used to control the translation between the internal number representation and the external format, are then explained. Finally, the topics of unformatted (or binary) I/O and direct-access files are covered.

## 9.2 Number conversion

The ways in which numbers are stored internally by a computer are the concern of neither the Fortran standard nor this book. However, if we wish to output values — to display them on a terminal or to print them — then their internal representations must be converted into a character string which can be read in a normal way. For instance, the contents of a given computer word may be BE1D7DBF (hexadecimal) and correspond to the value −0.000450 . For our particular purpose, we may wish to display this quantity as −.000450 , or as −4.5E−04 or rounded to one significant digit as −5E−04 . The conversion from the internal representation to the external form is carried out according to the information specified by an edit descriptor contained in a *format specification*. These will both be dealt with fully later in this chapter; for the moment, it is sufficient to give a few examples. For instance, to print an integer value in a field of 10 characters width, we would use the edit descriptor I10, where I stands for integer conversion, and 10 specifies the width of the output field. To print a real quantity in a field of 10 characters, five of which are reserved for the fractional part of the number, we specify F10.5 . F stands for floating-point (real) conversion, 10 is the total width of the output field and 5 is the width of the fractional part of the field. If the number given above were to be converted according to this edit descriptor, it would appear as *bb*−0.00045 , where *b* represents a blank. To print a character variable in a field of 10 characters, we would specify A10, where A stands for alphanumeric conversion.

A format specification consists of a list of edit descriptors enclosed in parentheses, and can be coded either as a default character expression, for instance

```
'(I10, F10.3, A10)'
```

or as a separate FORMAT statement, referenced by a statement label, for example

```
10 FORMAT(I10, F10.3, A10)
```

To print the scalar variables J, B, and C, of types integer, real, and character respectively, we may then write either

```
PRINT '(I10, F10.3, A10)', J,B,C
```

or

```
PRINT 10, J,B,C
10 FORMAT(I10, F10.3, A10)
```

The first form is normally used when there is only a single reference in a scoping unit to a given format specification, the second when there are several or when the format is complicated. The part of the statement designating the quantities to be printed is known as the *output list* and forms the subject of the following subsection.

## 9.3 I/O lists

The quantities to be read or written by a program are specified in an I/O list. For output, they may be expressions but for input must be variables. In both cases, list items may be implied-DO lists of quantities. Examples are shown in Figure 38, where we note the use of a *repeat count* in front of those edit descriptors that are required repeatedly.

```
INTEGER I
REAL, DIMENSION(10) :: A
CHARACTER(LEN=20) WORD
PRINT '(I10)',     I
PRINT '(10F10.3)', A
PRINT '(3F10.3)',  A(1),A(2),A(3)
PRINT '(A10)',     WORD(5:14)
PRINT '(5F10.3)', (A(I), I=1,9,2)
PRINT '(2F10.3)',  A(1)*A(2)+I, SQRT(A(3))
```

Figure 38.

In all these examples, except the last one, the expressions consist of single variables and would be equally valid in input statements using the READ statement, for example

```
READ '(I10)', I
```

Such statements may be used to read values which are then assigned to the variables in the input list.

If an array appears as an item, it is treated as if the elements were specified in array element order. For example, the third of the PRINT statements in Figure 38 could have been written

```
PRINT '(3F10.3)', A(1:3)
```

However, no element of the array may appear more than once. Thus, the case in Figure 39 is not allowed.

```
INTEGER J(10), K(3)
 :
K = (/ 1, 2, 1 /)
READ '(3I10)', J(K)     ! Illegal because J(1) appears twice
```

Figure 39.

Any pointers in an I/O list must be associated with a target, and transfer takes place between the file and the targets. An item of derived type is treated as if the components were specified in the same order as in the type declaration. This rule is applied repeatedly for components of derived type, so that it is as if we specified the list of items of intrinsic type that constitute its ultimate components. For example, if P and T are of the types POINT and TRIANGLE of Figure 1, the statement

```
READ '(8F10.5)', P, T
```

has the same effect as the statement

```
READ '(8F10.5)', P%X, P%Y, T%A%X, T%A%Y, T%B%X,       &
                 T%B%Y, T%C%X, T%C%Y
```

Each ultimate component must be accessible (not, for example, be a PRIVATE component of a PUBLIC type).

An object in an I/O list is not permitted to be of a derived type that has a pointer component at any level of component selection. One reason for this restriction is because of the problems associated with recursive data structures. For example, if CHAIN is a data object of the type ENTRY of Figure 3 (Section 2.14) and is set up to hold a chain of length three, it has as its ultimate components CHAIN%INDEX, CHAIN%NEXT%INDEX, CHAIN%NEXT%NEXT%INDEX, and CHAIN%NEXT%NEXT%NEXT, the last of which is a disassociated pointer.

The general form of an implied-DO list is

(do-object-list, do-var = expr, expr[,expr])

where each *do-object* is a variable (for input), an expression (for output), or is itself an implied-DO list; *do-var* is a named scalar integer variable; and each *expr* is a scalar integer expression. The loop initialization and execution is the same as for a (possibly nested) set of DO constructs (Section 4.5). In an input list, a variable that is an item in a *do-object-list* must not be a *do-var* of any implied-DO list in which it is contained, nor be associated[14] with such a *do-var*. In an input or output list, no *do-var* may be a *do-var* of any implied-DO list in which it is contained or be associated with such a *do-var*.

Note that a zero-sized array, or an implied-DO list with a zero iteration count, may occur as an item in an I/O list. Such an item corresponds to no actual data transfer.

---

[14] Such an illegal association could be established by pointer association.

## 9.4  Format definition

In the PRINT and READ statements of the previous subsection, the format specification was given each time in the form of a character constant immediately following the keyword.  In fact, there are three ways in which a format specification may be given.  They are:

i) As a statement label referring to a FORMAT statement containing the relevant specification between parentheses:

```
      PRINT 100, Q
      :
  100 FORMAT(F10.3)
```

The FORMAT statement must appear in the same scoping unit, before the CONTAINS statement if it has one.  It is customary either to place each FORMAT statement immediately after the first statement which references it, or to group them all together just before the CONTAINS or END statement.  It is also customary to have a separate sequence of numbers for the statement labels used for FORMAT statements.  A given FORMAT statement may be used by any number of formatted I/O statements, whether for input or for output.

ii) As a default character expression whose value commences with a format specification in parentheses:

```
      CHARACTER(LEN=*), PARAMETER :: FORM='(F10.3)'
      :
      PRINT FORM, Q
```

or

```
      CHARACTER CARRAY(7) =  (/ '(','F','1','0','.','3',')' /)
      :
      PRINT CARRAY, Q ! Elements of an array expression
                      ! are concatenated
```

or

```
CHARACTER(4) CARR1(10)
CHARACTER(3) CARR2(10)
INTEGER I, J
  :
CARR1(10) = '(F10'
CARR2(3) = '.3)'
  :
I = 10
J = 3
  :
PRINT CARR1(I)//CARR2(J), Q
```

or, simply,

```
PRINT '(F10.3)', Q
```

From these examples it may be seen that it is possible to program formats in a flexible way, and particularly that it is possible to use arrays, expressions and also substrings in a way which allows a given format to be built up dynamically at execution-time from various components. Any character data which might follow the trailing right parenthesis are ignored and may be undefined. In the case of an array, its elements are concatenated in array element order. However, on input *no* component of the format specification may appear also in the input list, or be associated with it. This is because the standard requires that the whole format specification be established *before* any I/O takes place. Further, no redefintion or undefinition of any characters of the format is permitted during the execution of the I/O statement.

iii) As an asterisk. This is a type of I/O known as *list-directed* I/O, in which the format is defined by the computer system at the moment the statement is executed, depending on both the type and magnitude of the entities involved. This facility is particularly useful for the input and output of small quantities of values, especially in temporary code which is used for test purposes, and which is removed from the final version of the program:

```
PRINT *, 'Square-root of Q = ', SQRT(Q)
```

This example outputs a character constant describing the expression which is to be output, followed by the value of the expression under investigation. On the terminal screen, this might appear as

```
Square-root of Q = 4.392246
```

the exact format being dependent on the computer system used. Character strings in this form of output are normally undelimited, as if an A edit descriptor were in use, but an option in the OPEN statement (Section 10.3) may be used to require that they be delimited by apostrophes or quotation marks. Except for undelimited strings, values are separated by spaces or commas. Logical variables are represented as T for true and F for false. The processor may represent a sequence of $r$ identical values $c$ by the form $r*c$. Further details of list-directed input/output are deferred until Section 9.9.

Blank characters may precede the left parenthesis of a format specification, and may appear at any point within a format specification with no effect on the interpretation, except within a character string edit descriptor (Section 9.13.3).

## 9.5  Unit numbers

Input/output operations are used to transfer data between the variables of an executing program, as stored in the computer, and an external medium. There are many types of external media: the terminal, printer, disc drive, and magnetic cartridge are perhaps the most familiar. Whatever the device, a Fortran program regards each one from which it reads or to which it writes as a *unit*, and each unit, with two exceptions, has associated with it a *unit number*. This number must not be negative and is often in the range 1 to 99. Thus we might associate with a disc drive from which we are reading the unit number 10, and to a magnetic cartridge drive to which we are writing the unit number 11. All program units of an executable program that refer to a particular unit number are referencing the same file.

There are two I/O statements, PRINT and a variant of READ, that do not reference any unit number; these are the statements that we have used so far in examples, for the sake of simplicity. A READ statement without a unit number will normally expect to read from the terminal, unless the program is working in batch (non-interactive) mode in which case there will be a disc file with a reserved name from which it reads. A PRINT statement will normally expect to output to the terminal, unless the program is in batch mode in which case another disc file with a reserved name will be used. Such files are usually suitable for subsequent output on a physical output device. The system may implicitly associate unit numbers to these default units.

Apart from these two special cases, all I/O statements must refer explicitly to a unit in order to identify the device to which or from which data are to be transferred. The unit may be given in one of three forms. These are shown in the following  examples which use another form of the READ containing a unit specifier, *u*, and format specifier, *fmt*, in parentheses and separated by a comma, where *fmt* is a format specification as described in the previous subsection:

　　READ (*u*, *fmt*) *list*

The three forms of *u* are:

i) As a scalar integer expression that gives the unit number:

```
READ (4, '(F10.3)') Q
READ (NUNIT, '(F10.3)') Q
READ (4*I+J, 100) A
```

where the value may be any nonnegative integer allowed by the system for this purpose.

ii) As an asterisk:

```
READ (*, '(F10.3)') Q
```

where the asterisk implies the standard input unit designated by the system, the same as that used for READ without a unit number.

iii) As a default character variable identifying an *internal file* (see next subsection).

## 9.6  Internal files

Internal files allow format conversion between various representations to be carried out by the program in a storage area defined within the program itself. There are two particularly useful applications, one to read data whose format is not properly known in advance, and the other to prepare output lists containing mixed character and numerical data, all of which has to be prepared in character form, perhaps to be displayed as a caption on a graphics display. The character data must be of default kind. The first application will now be described; the second will be dealt with in Section 9.8.

　　Imagine that we have to read a string of 30 digits, which might correspond to 30 one-digit integers, 15 two-digit integers or 10 three-digit integers. The information as to which type of data is involved is given by the value of an additional digit, which has the value 1, 2, or 3, depending on the number of digits each integer contains. An internal file provides us with a mechanism whereby the 30 digits can be read into a character buffer area. The value of the final digit can be tested separately, and 30, 15, or 10 values read from the internal file, depending on this value. The basic code to achieve this might read as follows (no error recovery or data validation is included, for simplicity):

```
INTEGER       IVAL(30), KEY, I
CHARACTER(30) BUFFER
CHARACTER(6)  FORM(3)= (/ '(30I1)', '(20I2)', '(10I3)' /)
READ (*, '(A30,I1)')      BUFFER, KEY
READ (BUFFER, FORM (KEY)) (IVAL(I), I=1,30/KEY)
```

IVAL is an array which will receive the values, BUFFER a character variable of a length sufficient to contain the 30 input digits, and FORM a character array containing the three possible formats to which the input data might correspond. The first READ statement reads 30 digits into BUFFER as character data, and a final digit into the integer variable KEY. The second READ statement reads the data from BUFFER into IVAL, using the appropriate conversion as specified by the edit descriptor selected by KEY. The number of variables read from BUFFER to IVAL is defined by the implied-DO loop, whose second specifier is an integer expression depending also on KEY. After execution of this code, IVAL will contain 30/KEY values, their number and exact format not having been known in advance.

If an internal file is a scalar, it has a single record whose length is that of the scalar. If it is an array, its elements, in array element order, are treated as successive records of the file and each has length that of an array element. It may not be an array section with a vector subscript.

A record becomes defined when it is written. The number of characters sent must not exceed the length of the record. It may be less, in which case the rest of the record is padded with blanks. For list-directed output (Section 9.4), character constants are not delimited. A record may be read only if it is defined (which need not only be by an output statement). Records are padded with blanks, if necessary.

An internal file is always positioned at the beginning of its first record prior to data transfer (the array section notation may be used to start elsewhere in an array). Of course, if an internal file is an allocatable array or pointer, it must be allocated or associated with a target. Also, no item in the input/output list may be in the file or associated with the file.

## 9.7 Formatted input

In the previous sections we have given complete descriptions of the ways that formats and units may be specified, using simplified forms of the READ and PRINT statements as examples. There are, in fact, two forms of the formatted READ statement. Without a unit, it has the form

READ *fmt* [,*list*]

and with a unit it may take the form

READ ([UNIT=]*u*, [FMT=]*fmt* [, IOSTAT=*ios*]            &
[, ERR=*error-label*] [, END=*end-label*]) [*list*]

where *u* and *fmt* are the unit and format specifiers described in the previous two subsections; IOSTAT=, ERR=, and END= are optional specifiers which allow a user to specify how a READ statement shall recover from various exceptional conditions; and *list* is a list of variables and implied-DO lists of variables. The keyword items may be specified in any order, although it is usual to keep the unit number and format specification as the first two. The unit number must be first if it does not have its keyword. If the format does not have its keyword, it must be second, following the unit number without its keyword. Note that this parallels the rules for keyword calls of procedures, except that the positional list is limited to two items.

For simplicity of exposition, we have so far limited ourselves to formats that correspond to a single record in the file, but we will meet later in this chapter cases that lead to the input of a part of a record or of several successive records.

The meanings of the optional specifiers are as follows. If the IOSTAT= is specified, then *ios* must be a scalar integer variable of default kind which, after execution of the READ statement, has a negative value if an end-of-record condition is encountered during non-advancing input (Section 9.12), a different negative value if an endfile condition was detected on the input device (Section 10.2.3), a positive value if an error was detected (for instance a parity error), or the value zero otherwise. The actual values assigned to *ios* in the event of an exception occurring are not defined by the standard, only the signs.

If the END= is specified, then *end-label* must be a statement label of a statement in the same scoping unit, to which control will be transferred in the event of the end of the file being reached. For an external file, the file is positioned after the endfile record.

If the ERR= is specified, then *error-label* is a statement label in the same scoping unit, to which control will be transferred in the event of any other exception occurring. The labels *error-label* and *end-label* may be the same. If they are not specified and an exception occurs, execution will stop, unless IOSTAT is specified. An example of a READ statement with its associated error recovery is given in Figure 40, in which ERROR and LAST_FILE are subroutines to deal with the exceptions. They will normally be system dependent.

If an error or end-of-file condition occurs on input, the file position becomes indeterminate and all list items become undefined. If an end-of-file condition occurs for an external file, the file is positioned following the endfile record (Section 10.2.3); if there is otherwise an error condition, the file position is indeterminate.  An end-of-file condition occurs also if an attempt is made to read beyond the end of an internal file.

```
      READ (NUNIT, 210, IOSTAT=IOS, ERR=110, END=120) A,B,C
  !
  !   Successful read - continue execution
      :
      :
  !
  !   Error condition - take appropriate action
110   CALL ERROR (IOS)
      GO TO 999
  !
  !   END-OF-FILE condition - test whether more
  !   files follow
120   CALL LAST_FILE
      :
210   FORMAT(3F10.3)
999   END
```

Figure 40.

It is a good practice to include some sort of error recovery in all READ statements which are included permanently in a program.  On the other hand, input for test purposes is normally sufficiently well handled by the simple form of READ without unit number, and without error recovery.

## 9.8  Formatted output

There are two types of formatted output statements, the PRINT statement which has appeared in many of the examples so far in this chapter, and the WRITE statement whose syntax is similar to that of the READ statement:

PRINT *fmt* [,*list*]

and

WRITE ([UNIT=] *u*, [FMT=]*fmt* [,IOSTAT=*ios*]        &
        [,ERR=*error-label*] ) [*list*]

where all the components have the same meanings as described for the READ statement (Section 9.7).  An asterisk for *u* specifies the standard output unit, as used by PRINT.  If an error or end-of-file condition occurs on output, execution of the statement terminates, any implied-DO variables become undefined, and the file position becomes indeterminate.

An example of a WRITE statement is

```
    WRITE (NOUT, 100, IOSTAT=IOS, ERR=110) A
    :
100  FORMAT(10F10.3)
```

An example using an internal file is given in Figure 41, which builds a character string from numeric and character components. The final character string might be passed to another subroutine for output, for instance as a caption on a graphics display.

---

```
INTEGER DAY
REAL CASH
CHARACTER(LEN=50) LINE
:
!   Write into line
WRITE (LINE,'(A, I2, A, F8.2, A)')                          &
    'Takings for day ', DAY, ' are ', CASH, ' dollars'
```

---

Figure 41.

In this example, we declare a character variable that is long enough to contain the text to be transferred to it. (The WRITE statement contains a format specification with A edit descriptors without a field width. These assume a field width corresponding to the actual length of the character strings to be converted.) After execution of the WRITE statement, LINE might contain the character string

```
    Takings for day  3 are   4329.15 dollars
```

and this could be used as a string for further processing.

The number of characters written to LINE must not exceed its length.

## 9.9  List-directed input/output

In Section 9.3, the list-directed output facility using an asterisk as format specifier was introduced. We assumed that the list was short enough to fit into a single record, but for long lists the processor is free to output several records. Delimited character constants may be split between records, and complex constants that are as long as, or longer than, a record may be split before or after the comma that separates the two parts. Apart from these cases, a value always lies within a single record. For the sake of carriage control (see Section 9.10), the first character of each record is blank unless a delimited character constant is being continued. Note that when an undelimited character constant is continued, the first character of the continuation record is blank. The only blanks permitted in a numeric constant are within a split complex constant after the comma.

This facility is equally useful for input, especially of small quantities of test data. On the input record, the various constants may appear in any of their usual forms, just as if they were being read under the usual edit descriptors, as defined in Section 9.13. Exceptions are that complex values must be enclosed in parentheses, character constants may be delimited, a blank must not occur except in a delimited character constant or in a complex constant before or after a numeric field, blanks are never interpreted as zeros, and the optional characters which are allowed in a logical constant (those other than T and F, see Section 9.13.2) must include neither a comma nor a slash.

Character constants that are enclosed in apostrophes or quotation marks may be spread over as many records as necessary to contain them, except that a doubled quotation mark or apostrophe must not be split between records. Delimiters may be omitted for a default character constant if

- the constant does not contain a blank, comma, or slash;

- it is contained in one record;

- the first character is neither a quotation mark nor an apostrophe; and

- the leading characters are not numeric followed by an asterisk.

In this case, the constant is terminated when a blank, comma, slash, or end of record is encounterd, and apostrophes or quotation marks appearing within the constant must not be doubled.

Whenever a character value has a different length from the corresponding list item, the value is truncated or padded on the right with blanks, as in the character assignment statement.

It is possible to use a repeat count for a given constant, for example 6*10 to specify six occurrences of the integer value 10.

The (optionally repeated) constants are separated in the input by *separators*. A separator is one of the following, appearing other than in a character constant:

- a comma, optionally preceded and optionally followed by one or more contiguous blanks,

- a slash (/), optionally preceded and optionally followed by one or more contiguous blanks, or

- one or more contiguous blanks between two non-blank values or following the last non-blank value.

An end of record not within a character constant is regarded as a blank and, therefore, forms part of a separator. A blank embedded in a complex constant or delimited character constant is not a separator. An input record may be terminated by a slash separator, in which case all the following values in the record are ignored, and the input statement terminates.

If there are no values between two successive separators, or between the beginning of the first record and the first separator, this is taken to represent a *null value* and the corresponding item in the input list is left unchanged, defined or undefined as the case may be. A null value must not be used for the real or imaginary part of a complex constant, but a single null value may be used for the whole complex value. A series of null values may be represented by a repeat count without a constant: ,6*, . When a slash separator is encountered, null values are given to any remaining list items.

Examples of this form of the READ statement are:

```
INTEGER I
REAL A
COMPLEX FIELD(2)
LOGICAL FLAG
CHARACTER (LEN=12) TITLE
CHARACTER (LEN=4) WORD
  :
READ *, I, A, FIELD, FLAG, TITLE, WORD
```

If this reads the input record

```
10b6.4b(1.,0.)b(2.,0.)bTbTEST
```

(in which *b* stands for a blank, and blanks are used as separators), then I, A, FIELD, FLAG, and TITLE will acquire the values 10, 6.4, (1.,0.) and (2.,0.), .TRUE. and TEST respectively. WORD remains unchanged. For the input records

```
10,.64E1,2*,.TRUE.
'HISTOGRAMb10'/VAL1
```

(in which commas are used as separators), the variables I, A, FLAG, and TITLE will acquire the values 10, 6.4, .TRUE., and HISTOGRAM*b*10 respectively. FIELD and WORD remain unchanged, and the input string VAL1 is ignored as it follows a slash. (Note the apostrophes, which are required as the string contains a blank. Without delimiters, this string would appear to be a string followed by the integer value 10.) Because of this slash, the read statement does not continue with the next record and the list is thus not fully satisfied.

## 9.10 NAMELIST

It can be useful, especially for program testing, to input or output an annotated list of values. The values required are specified in a NAMELIST group (Section 7.15), and the I/O is performed by a READ or WRITE statement that does not have an I/O list, and in which either the format is replaced by a namelist-group name as the second positional parameter or the FMT= specifier is replaced by a NML= specifier with that name. When reading, only those objects which are specified in the input record and which do not have a null value become defined. All other list items remain in their existing state of definition or undefinition. It is possible to define the value of an array element or section without affecting the other portions of the array. When writing, all the items in the group are written to the file specified. This form of I/O is not available for internal files.

The value for a scalar object or list of values for an array is preceded in the records by the name or designator and an equals sign which may optionally be preceded or followed by blanks. The form of the list of values and null values in the input and output records is as that for list-directed I/O (Section 9.9), except that character constants must *always* be delimited. A NAMELIST input statement terminates when a name-value sequence has been processed for every name in the NAMELIST group or on the appearance of a slash in the list outside a character constant. A simple example is

```
INTEGER      NO_OF_EGGS, LITRES_OF_MILK, KILOS_OF_BUTTER
NAMELIST/FOOD/NO_OF_EGGS, LITRES_OF_MILK, KILOS_OF_BUTTER
READ (5, NML=FOOD)
```

to read the record

```
&FOOD LITRES_OF_MILK=5, NO_OF_EGGS=12 /
```

where we note that the order of the two values given is not the same as their order in the NAMELIST group — the orders need not necessarily match. The value of KILOS_OF_BUTTER remains unchanged. The first non-blank item in the record is an ampersand followed without an intervening blank by the group name. The slash is obligatory as a terminator. On output, a similar annotated list of values is produced, starting with the name of the group and ending with a slash. Here the order is that of the NAMELIST group. Thus, the statements

```
INTEGER NUMBER, LIST(10)
NAMELIST/OUT/NUMBER, LIST
WRITE (6, NML=OUT)
```

might produce the record

```
&OUT NUMBER=1, LIST=14, 9*0 /
```

On output, the names are always in upper case.  No output is produced by a zero-sized object.

Where a subobject designator appears in an input record, all substring expressions, subscripts, and strides must be integer literal constants.  All group names, object names, and component names are interpreted without regard to case.  Blanks may precede or follow the name or designator, but must not appear within it.

If the object is scalar and of intrinsic type, the equals sign must be followed by one value.  If it is of derived type and is an array, the equals sign must be followed by a list of values of intrinsic type corresponding to the replacement of each derived-type value by its ultimate components, and each array by its elements in array element order.

The list of values must not be too long, but it may be too short, in which case trailing null values are regarded as having been appended.  If an object is of type character, the corresponding item must be of the same kind.

Zero-sized objects must not appear in a NAMELIST input record.  In any multiple occurrence of an object in an input record, the final value is taken.

## 9.11  Carriage control

Fortran's formatted output statements were originally designed for line-printers, with their concept of lines and pages of output.  On such a device, the first character of each output record must be of default kind.  It is not printed but interpreted as a *carriage control character*.  If it is a blank, no action is taken, and it is good practice to insert a blank as the first character of each record, either explicitly as ' ' or using the T2 edit descriptor (described in the next section), in order to avoid inadvertent generation of spurious carriage control characters.  This can happen when the first character in an output record is non-blank, and might occur, for instance, when printing integer values with the format '(I5)'.  Here all output values between −999 and 9999 will have a blank in the first position, but all others will generate a character there which may be used mistakenly for carriage control.

The carriage control characters defined by the standard are:

| | |
|---|---|
| *b* | to start a new line |
| + | to remain on the same line (overprint) |
| 0 | to skip a line |
| 1 | to advance to the beginning of the next page |

As a precaution, the first character of each record produced by list-directed and NAMELIST output is a blank, unless it is the continuation of a delimited character constant.

In this context, we note that execution of a PRINT statement does not imply that any printing will actually occur, and nor does execution of a WRITE statement imply that printing will not occur.

## 9.12  Non-advancing I/O

So far we have considered each READ or WRITE statement to perform the input or output of a complete record. There are, however, many applications, especially in screen management, where this would become an irksome restriction. What is required is the ability to read and write without always advancing the file position to ahead of the next record. This facility is provided by *non-advancing* I/O. To gain access to this facility, the optional ADVANCE= specifier must appear in the READ or WRITE statement and be associated with a scalar default character expression *advance* which evaluates, after suppression of any trailing blanks and conversion of any lower-case letters to upper case, to the value NO. The only other allowed value is YES which is the default value if the specifier is absent; in this case normal (advancing) I/O occurs.

The following optional specifiers are available for a non-advancing READ statement:

> EOR=*eor-label*
> SIZE= *size*

where *eor-label* is a statement label in the same scoping unit and *size* is a default integer scalar variable. The *eor-label* may be the same as an *end-label* or *error-label* of the READ statement.

An advancing I/O statement always repositions the file after the last record accessed, but a non-advancing I/O statement performs no such repositioning and may therefore leave the file positioned within a record. If a non-advancing input statement attempts to transfer data from beyond the end of the *current* record, an end-of-record condition occurs. The IOSTAT variable, if present, will acquire a different negative value to the one indicating an end-of-file condition; and, if the EOR= specifier is present, control is transferred to the statement specified by its associated *eor-label*. In order to provide a means of controlling this process, the SIZE= specifier, when present, sets *size* to the number of characters actually read. A full example is thus

```
      CHARACTER(3) KEY
      INTEGER UNIT, SIZE
      READ (UNIT, '(A3)', ADVANCE='NO', SIZE=SIZE, EOR=66) KEY
      :
   ! KEY is not in one record
66   KEY = ''
      :
```

As for error and end-of-file conditions, the program terminates when an end-of-record condition occurs if neither EOR= nor IOSTAT= is specified.

If encountering an end-of-record on reading results in the input list not being satisfied, the PAD= specifier described in Section 10.3 will determine whether any padding with blank characters occurs. Blanks inserted as padding are not included in the SIZE= count.

It is possible to perform normal and non-advancing I/O on the same record or file. For instance, a non-advancing read might read the first few characters of a record and a normal read the remainder.

A particular application of this facility is to write a prompt to a terminal screen and to read from the next character position on the screen without an intervening line-feed:

```
      WRITE (*, *, ADVANCE='NO') 'Enter next prime number:'
      READ  (*, '(I10)') PRIME_NUMBER
```

Non-advancing I/O may be performed only on an external file, and may not be used for NAMELIST or list-directed I/O. Note that, as for advancing input/output, several records may be processed by a single statement.

## 9.13  Edit descriptors

In the description of the possible forms of a format specification in Section 9.4, a few examples of the edit descriptors were given. As mentioned there, edit descriptors give a precise specification of how values are to be converted into a character string on an output device or internal file, or converted from a character string on an input device or internal file to internal representations.

With certain exceptions noted in the following text, edit descriptors in a list are separated by commas, and only in the case where an input/output list is empty or specifies only zero-length objects may there be no edit descriptor at all in the format specification. On a processor that supports upper- and lower-case letters, edit descriptors are interpreted without regard to case.

## 9.13.1  Repeat counts

Edit descriptors fall into three classes: *data, control,* and *character-string.* The data edit descriptors may be preceded by a repeat count (an unsigned default integer literal constant), as in the example

```
10F12.3
```

Of the remaining edit descriptors, only the slash edit descriptor (Section 9.13.4) may have an associated repeat count.  A repeat count may be applied to a group of edit descriptors, enclosed in parentheses:

```
PRINT '(4(I5,F8.2))', (I(J), A(J), J=1,4)
```

(for integer I and real A).  This is equivalent to writing

```
PRINT '(I5,F8.2,I5,F8.2,I5,F8.2,I5,F8.2)', (I(J), A(J), J=1,4)
```

Repeat counts such as this may be nested:

```
PRINT '(2(2I5,2F8.2))', I(1),I(2),A(1),A(2),I(3),I(4),A(3),A(4)
```

If a format specification without components in parentheses is used with an I/O list that contains more elements than the number of edit descriptors, taking account of repeat counts, then a new record will begin, and the format specification repeated.  Further records begin in the same way until the list is exhausted.  To print an array of 100 integer elements, 10 elements to a line, the following statement might be used:

```
PRINT '(10I8)', (I(J), J=1,100)
```

Similarly, when reading from an input file, new records would be read until the list is satisfied, a new record being taken from the input file each time the specification is repeated *even if the individual records contain more input data than specified by the format specification.*  These superfluous data would be ignored.  For example, reading the two records (*b* again stands for a blank)

```
bbb10bbb15bbb20
bbb25bbb30bbb35
```

under control of the READ statement

```
READ '(2I5)', I,J,K,L
```

would result in the four integer variables I, J, K and L acquiring the values 10, 15, 25 and 30, respectively.

If a format contains components in parentheses, as in

```
100    FORMAT(2I5, 3(I2,2(I1,I3)), 2(2F8.2,I2))
```

whenever the format is exhausted, a new record is taken and format control reverts to the repeat factor preceding the left parenthesis corresponding to the last-but-one right parenthesis, here 2(2F8.2,I2), or to the parenthesis itself if it has no repeat factor. This we call *reversion*.

## 9.13.2 Data edit descriptors

Values of all the intrinsic data types may be converted by the G edit descriptor. However, for reasons of clarity, it is described last.

*Integer* values may be converted by means of the I edit descriptor. This comes in a basic form, I$w$, which defines the width of a field, $w$, a non-zero unsigned default integer literal constant. The integer value will be read from or written to this field, adjusted to its right-hand side. If we again designate a blank position by $b$ then the value −99 printed under control of the edit descriptor I5 will appear as $bb$−99, the sign counting as one position in the field.

For output, an alternative form of this edit descriptor allows the number of digits which are to be printed to be specified exactly, even if some are leading zeros. The form I$w.m$ specifies the width of the field, $w$, and that at least $m$ digits are to be output, where $m$ is an unsigned default integer literal constant. The value 99 printed under control of the edit descriptor I5.3 would appear as $bb$099. On input, I$w.m$ is interpreted in exactly the same way as I$w$.

For the I and all other numeric edit descriptors, if the output field is too narrow to contain the number to be output, it is filled with asterisks.

Integer values may also be converted by the B$w$, B$w.m$, O$w$, O$w.m$, Z$w$, and Z$w.m$ edit descriptors. These are similar to the I form, but are intended for integers represented in the binary, octal, and hexadecimal number systems, respectively (Section 2.6.1). The external form does not contain the leading letter (B, O, or Z) or the delimiters.

*Real* values may be converted by either E, EN, ES, or F edit descriptors. The F descriptor we have met in earlier examples. Its general form is F$w.d$, where $w$ and $d$ are unsigned default integer literal constants which define, respectively, the field width and the number of digits to appear after the decimal point in the output field. The decimal point counts as one position in the field.  On input, if the input string has a decimal point, the value of $d$ is ignored.  Reading the value $b9.3729b$ with the edit descriptor F8.3 would cause the value 9.3729 to be transferred.  All the digits are used, but roundoff may be inevitable because of the actual physical storage reserved for the value on the computer being used.

There are, in addition, two other forms of input string that are acceptable to the F edit descriptor.  The first is a signed string of digits without a decimal point.  In this case, the $d$ rightmost digits will be taken to be the fractional part of the value.  Thus $b-14629$ read under control of the edit descriptor F7.2 will transfer the value $-146.29$ .  The second form is the standard real form of literal constant, as defined in Section 2.6.2, and the variant in which the exponent is signed and E is omitted.  In this case, the $d$ part of the descriptor is again ignored.  Thus the value 14.629E-2 (or 14.629-2), under control of the edit descriptor F9.1, will transfer the value 0.14629.  The exponent letter may be written in lower case.

Values are rounded on output following the normal rules of arithmetic.  Thus, the value 10.9336, when output under control of the edit descriptor F8.3, will appear as $bb10.934$, and under the control of F4.0 as $b11.$ .

The E edit descriptor has two forms, E$w.d$ and E$w.d$E$e$, and is more appropriate for numbers with a magnitude below about 0.01, or above 1000.  The rules for these two forms for input are identical to those for the F$w.d$ edit descriptor.  For output with the E$w.d$ form of the descriptor, a different character string will be transferred, containing a significand with absolute value less than 1 and an exponent field of four characters that consists of either E followed by a sign and two digits or of a sign and three digits.  Thus, for $1.234 \times 10^{23}$ converted by the edit descriptor E10.4, the string $b.1234E+24$ or $b.1234+024$ will be transferred.  The form containing the exponent letter E is not used if the magnitude of the exponent exceeds 99.  For instance, E10.4 would cause the value $1.234 \times 10^{-150}$ to be transferred as $b.1234-149$ .  Some processors print a zero before the decimal point.

In the second form of the E edit descriptor, E$w.d$E$e$, $e$ is an unsigned, non-zero default integer literal constant that determines the number of digits to appear in the exponent field.  This form is obligatory for exponents whose magnitude is greater than 999.  Thus the value $1.234 \times 10^{1234}$ with the edit descriptor E12.4E4 is transferred as the string $b.1234E+1235$.  An increasing number of computers are able to deal with these very large exponent ranges.

The EN (*engineering*) edit descriptor is identical to the E edit descriptor except that on output the decimal exponent is divisible by three, a non-zero significand is greater than or equal to 1 and less than 1000, and the scale factor (Section 9.13.4) has no effect.  Thus, the value 0.00217 transferred under an EN9.3 edit descriptor would appear as 2.170E-03 or 2.170-003.

The ES (*scientific*) edit descriptor is identical to the E edit descriptor, except that on output the absolute value of a non-zero significand is greater than or equal to 1 and less than 10 and the scale factor (Section 9.13.4) has no effect.   Thus, the value 0.00217 transferred under an ES9.3 edit descriptor would appear as 2.170E–03 or 2.170–003.

*Complex* values may be edited under control of pairs of F, E, EN, or ES edit descriptors.   The two descriptors do not need to be identical.   The complex value (0.1,100.) converted under control of F6.1,E8.1 would appear as *bbb*0.1*b*0.1E+03 .   The two descriptors may be separated by character string and control edit descriptors (to be described in Sections 9.13.3 and 9.13.4.)

*Logical* values may be edited using the L*w* edit descriptor.   This defines a field of width *w* which on input consists of optional blanks, optionally followed by a decimal point, followed by T or F (or t or f), optionally followed by additional characters.   Thus a field defined by L7 permits the strings .TRUE. and .FALSE. to be input. The characters T or F will be transferred as the values true or false respectively.   On output, the character T or F will appear in the right-most position in the output field.

*Character* values may be edited using the A edit descriptor in one of its two forms, either A or A*w*.   In the first of the two forms, the width of the input or output field is determined by the actual width of the item in the I/O list, measured in number of characters of whatever kind.   Thus, a character variable of length 10, containing the value STATEMENTS, when written under control of the A edit descriptor would appear in a field 10 characters wide, and the non-default character variable of length 4 containing the value 国際標準 would appear in a field 4 characters wide.   If, however, the first variable were converted under an A11 edit descriptor, it would be printed with a leading blank, *b*STATEMENTS.   Under control of A8, the eight left-most characters only would be written: STATEMEN.

Conversely, with the same variable on input, an A11 edit descriptor would cause the 10 right-most characters in the 11 character-wide input field to be transferred:   *b*STATEMENTS would be transferred as STATEMENTS.   The A8 edit descriptor would cause the eight characters in the field to be transferred to the eight left-most positions in the variable, and the remaining two would be filled with blanks:   STATEMEN would be transferred as STATEMEN*bb*.

All characters transferred under the control of an A or A*w* edit descriptor must be of the same kind as the I/O list item, and we note that this edit descriptor is the *only* one which can be used to transmit non-default characters to or from a record.   In the non-default case, the blank padding character is processor dependent.

The G*w.d* and G*w.dEe* (*general*) edit descriptor may be used for any intrinsic data type. When used for real or complex types, it is identical to the E edit descriptor except that output values zero and with magnitude N in the range $0.1-0.5\times10^{-d-1} \leq N < 10^d-0.5$ are converted as if by an F edit descriptor, and followed by the same number of blanks as the E edit descriptor would have used for the exponent part. The equivalent F edit descriptor is F*w'.d'*, where $w' = w-4$ for G*w.d* or $w-e-2$ for G*w.dEe*, and $d' = d-k$ when N lies in the range $10^{k-1}(1-0.5\times10^{-d}) \leq N < 10^k(1-0.5\times10^{-d})$ for $k = 0,1,...,d$ and $d' = d-1$ when $N = 0$. This form is useful for printing values whose magnitudes are not well known in advance, and where an F conversion is preferred where possible, and an E otherwise.

When the G edit descriptor is used for integer, logical, or character types, it follows the rules of the I*w*, L*w*, and A*w* edit descriptors, respectively.

Finally, values of ***derived types*** are edited by the appropriate sequence of edit descriptors corresponding to the intrinsic types of the ultimate components of the derived type. An example is:

```
TYPE STRING
   INTEGER LENGTH
   CHARACTER(LEN=20) WORD
END TYPE STRING
TYPE(STRING) :: TEXT
READ(*, '(I2, A)') TEXT
```

## 9.13.3  Character string edit descriptor

A *default character* literal constant can be transferred to an output file by embedding it in the format specification itself, as in the example

```
100  FORMAT(' This is a FORMAT statement')
```

The string will appear each time a

```
PRINT 100
```

statement is executed. In this descriptor, case is significant. Character string edit descriptors must not be used on input.

## 9.13.4 Control edit descriptors

It is sometimes necessary to give other instructions to an I/O device than just the width of fields and how the contents of these fields are to be interpreted. For instance, it may be that one wishes to position fields at certain columns or to start a new record without issuing a new WRITE command. For this type of purpose, the control edit descriptors provide a means of informing the processor which action has to be taken. Some of these edit descriptors contain information that is used as it is processed; others are like switches, which change the conditions under which I/O takes place from the point where they are encountered, until the end of the processing of the I/O statement containing them (including reversions, Section 9.13.1). These latter descriptors we shall deal with first.

## Control edit descriptors setting conditions

*Embedded blanks* in numeric input fields are treated in one of two ways, either as zero, or as null characters which are squeezed out by moving the other characters in the input field to the right, and adding leading blanks to the field (unless the field is totally blank, in which case it is interpreted as zero). The default is given by the BLANK= specifier (Section 10.3) currently in effect for the unit or is nulls for an internal file. Whatever the default may then be for a file, it may be overridden during a given format conversion by the BN (blanks null) and BZ (blanks zero) edit descriptors. Let us suppose that the mode is that blanks are treated as zerosThe input string *bb1b4* converted by the edit descriptor I5 would transfer the value 104. The same string converted by BN,I5 would give 14. A BN or BZ edit descriptor switches the mode for the rest of that format specification, or until another BN or BZ edit descriptor is met. The BN and BZ edit descriptors have no effect on output.

Negative numerical values are always written with *leading signs* on output. For positive quantities other than exponents, whether the signs are written depends on the processor. The SS (sign suppress) edit descriptor suppresses leading plus signs, that is the value 99 printed by I5 is *bbb99* and 1.4 is printed by E10.2 as *bb0.14E+01* . To switch on plus sign printing, the SP (sign print) edit descriptors may be used: the same numbers written by SP,I5,E10.2 become *bb+99* and *b+0.14E+01*. The S edit descriptor restores the option to the processor. An SS, SP, or S will remain in force for the remainder of the format specification, unless another SS, SP, or S edit descriptor is met. These edit descriptors provide complete control over sign printing, and are useful for producing coded outputs which have to be compared automatically, on two different computers.

*Scale factors* apply to the input of real quantities under the E, F, EN, ES, and G edit descriptors, and are a means of scaling the input values. Their form is $k$P, where $k$ is a default integer literal constant specifying the scale factor. The value is zero at the beginning of execution of the statement. The effect is that any quantity which does not have an exponent field will be reduced by a factor $10^k$. Quantities with an exponent are not affected.

The scale factor $k$P also affects output with E, F or G editing, but has no effect with EN or ES editing. Under control of an F edit descriptor, the quantity will be multiplied by a factor $10^k$. Thus, the number 10.39 output by an F6.0 edit descriptor following the scale factor 2P will appear as $b$1039. . With the E edit descriptor, and with G where the E style editing is taken, the quantity is transferred with the exponent reduced by $k$, and the real part multiplied by $10^k$. Thus $0.31 \times 10^3$, written after a 2P edit descriptor under control of E9.2, will appear as 31.00E+01 . This gives a better control over the output style of real quantities which otherwise would have no significant digits before the decimal point.

The comma between a scale factor and an immediately following F, E, EN, ES, or G edit descriptor may be omitted, but we do not recommend that practice since it suggests that the scale factor applies only to the next edit descriptor, whereas in fact it applies throughout the format until another scale factor is encountered.

## Control edit descriptors for immediate processing

*Tabulation* in an input or output field can be achieved using the edit descriptors T$n$, TR$n$ (and $n$X), and TL$n$, where $n$ is a positive default integer literal constant. These state, respectively, that the next part of the I/O should begin at position $n$ in the current record (where the *left tab limit* is position 1), or at $n$ positions to the right of the current position, or at $n$ positions to the left of the current position (the left tab limit if the current position is less than or equal to $n$). Let us suppose that, following an advancing READ, we read an input record $bb$9876 with the following statement:

```
READ (*, '(T3, I4, TL4, I1, I2)') I, J, K
```

The format specification will move a notional pointer firstly to position 3, whence I will be read. I will acquire the value 9876, and the notional pointer is then at position 7. The edit descriptor TL4 moves it left four positions, back to position 3. The quantities J and K are then read, and they acquire the values 9 and 87, respectively. These edit descriptors cause replacement on output, or multiple reading of the same items in a record on input. On output, any gaps ahead of the last character actually written are filled with spaces. If any character that is skipped by one of the descriptors is of other than default type, the positioning is processor dependent.

If the current record is the first one processed by the I/O statement and follows non-advancing I/O that left the file positioned within a record, the next character is the left tab limit; otherwise, the first character of the record is the left tab limit.

The $n$X edit descriptor is equivalent to the TR$n$ edit descriptor. It is often used to place spaces in an output record.  For example, to start an output record with a blank by this method, one writes

```
100   FORMAT(1X,....)
```

Spaces such as this can precede a data edit descriptor, but 1X,I5 is not, for instance, exactly equivalent to I6 on output, as any value requiring the full six positions in the field will not have them available in the former case.

The T and X edit descriptors never cause replacement of a character already in an output record, but merely cause a change in the position within the record such that such a replacement might be caused by a subsequent edit descriptor.

*New records* may be started at any point in a format specification by means of the slash (/) edit descriptor.  This edit descriptor, although described here, may in fact have repeat counts; to skip, say, three records one can write either /// or 3/.  On input, a new record will be started each time a / is encountered, even if the contents of the current record have not all been transferred.  Reading the two records

```
bbb99bbb10
bb100bbb11
```

with the statement

```
READ '(BZ,I5,I3,/,I5,I3,I2)', I, J, K, L, M
```

will cause the values 99, 0, 100, 0 and 11 to be transferred to the five integer variables, respectively.  This edit descriptor does not need to be separated by a comma from a preceding edit descriptor, unless it has a repeat count; it does not ever need to be separated by a comma from a succeeding edit descriptor.

The result of writing with a format containing a sequence of, say, four slashes, as represented by

```
PRINT '(I5,4/,I5)', I, J
```

is to separate the two values by three blank records (the last slash starts the record containing J); if I and J have the values 99 and 100, they would appear as

```
bbb99
b
b
b
bb100
```

A slash edit descriptor written to an internal file will cause the following values to be written to the next element of the character array specified for the file. Each such element corresponds to a record, and the number of characters written to a record must not exceed its length.

*Colon editing* is a means of terminating format control if there are no further items in an I/O list. In particular, it is useful for preventing further output of character strings used for annotation if the output list is exhausted. Consider the following output statement, for an array L(3):

```
PRINT '("L1 = ", I5, :, "L2 = ", I5, :,"L3 = ", I5)', &
       (L(I) ,I=1,N)
```

If N has the value 3, then three values are printed. If N has the value 1 then, without the colons, the following output string would be printed:

```
L1 = 59 L2 =
```

The colon, however, stops the processing of the format, so that the annotation for the absent second value is not printed. This edit descriptor need not be separated from a neighbour by a comma. It has no effect if there are further items in the I/O list.

## 9.14 Unformatted I/O

The whole of this chapter has so far dealt with formatted I/O. The internal representation of a value may differ from the external form, which is always a character string contained in an input or output record. The use of formatted I/O involves an overhead for the conversion between the two forms, and often a roundoff error too. There is also the disadvantage that the external representation usually occupies more space on a storage medium than the internal representation. These three actual or potential drawbacks are all absent when unformatted I/O is used. In this form, the internal representation of a value is written exactly as it stands to the storage medium, and can be read back directly with neither roundoff nor conversion overhead. Here, a value of derived type is treated as a whole and is not equivalent to a list of its ultimate components. This is another reason for the rule (Section 9.3) that it must not have a pointer component at any level of pointer selection.

This type of I/O should be used in all cases where the records are generated by a program on one computer, to be read back on the same computer or another computer using the same internal number representations. Only when this is not the case, or when the data have to be visualized in one form or another, should formatted I/O be used. The records of a file must all be formatted or all be unformatted (apart from the endfile record).

Unformatted I/O has the incidental advantage of being simpler to program since no complicated format specifications are required. The forms of the READ and WRITE statements are the same as for formatted I/O, but without any FMT= or NML= specifier. Non-advancing I/O is not available (in fact, an ADVANCE= specifier is not allowed).

Each READ or WRITE statement transfers exactly one record. The file must be an external file. The number of values specified by the input list of a READ statement must not exceed the number of values available in the current record.

On output to a file connected for sequential access, a record of sufficient length is created. On input, the type and type parameters of each entity in the list must agree with those of the value in the record, except that two reals may correspond to one complex when all three have the same kind parameter.

## 9.15  Direct-access files

The only type of file organization that we have so far dealt with is the sequential file, which has a beginning and an end, and which contains a sequence of records, one after the other. Fortran permits another type of file organization known as *direct access* (or sometimes as random access or indexed). All the records have the same length, each record is identified by an index number, and it is possible to write, read, or re-write any specified record without regard to position. (In a sequential file, only the last record may be rewritten without losing other records; in general, records in sequential files cannot be replaced.) The records are either all formatted or all unformatted.

By default, any file used by a Fortran program is a sequential file, unless declared to be direct access. This declaration has to be made using the ACCESS='DIRECT' and RECL=$rl$ specifiers of the OPEN statement, which is described in the next chapter, ($rl$ is the length of a record in the file). Once this declaration has been made, reading and writing, whether formatted or unformatted, proceeds as described for sequential files, except for the addition of a REC= $i$ specifier to the READ and WRITE statements, where $i$ is a scalar integer expression whose value is the index number of the record concerned.

An END= specifier is not permitted. Usually, a data transfer statement for a direct-access file accesses a single record, but during formatted I/O any slash edit descriptor increases the record number by one and causes processing to continue at the beginning of this record. A sequence of statements to write, read, and replace a given record is given in Figure 42.

```
      PARAMETER (NUNIT=2, LEN=100)
      REAL A(LEN), B(LEN+1:2*LEN)
      :
      INQUIRE (IOLENGTH=LENGTH) A          ! See Section 10.5
      OPEN (NUNIT, ACCESS='DIRECT', RECL=LENGTH)
                                           ! See Section 10.3
      :
!  Write array B to direct-access file in record 14
      WRITE (NUNIT, REC=14) B
      :
!
!  Read the array back into array A
      READ (NUNIT, REC=14) A
      :
      DO I = 1, LEN/2
         A(I) = I
      END DO
!
!  Replace modified record
      WRITE (NUNIT, REC=14) A
```

Figure 42.

The file must be an external file and NAMELIST formatting, list-directed formatting, and non-advancing I/O are all unavailable.

Direct-access files are particularly useful for applications which involve lots of hopping around inside a file, or where records need to be replaced, for instance in data base applications. A weakness is that the length of all the records must be the same, though on formatted output, the record is padded with blanks if necessary. For formatted output, if the record is not filled, the remainder is undefined.

This simple and powerful facility allows much clearer control logic to be written than is the case for a sequential file which is repeatedly read, backspaced, or rewound. Only when direct-access files become large may problems of long access times become evident on some computer systems, and this point should always be investigated before heavy investments are made in programming large direct-access file applications.

Some computer systems allow the same file to be regarded as sequential or direct access according to the specification in the OPEN statement or its default. The standard, therefore, regards this as a property of the connection rather than of the file.  In this case, the order of records, even for sequential I/O, is that determined by the direct-access record numbering.

## 9.16  Execution of a data transfer statement

So far, we have used simple illustrations of data transfer statements without dependencies.  However, some forms of dependency are permitted and can be very useful.  For example, the statement

```
READ (*, *)N, A(1:N)                  ! N is an integer
```

allows the length of an array section to be part of the data.

With dependencies in mind, the order in which operations are executed is important.  It is as follows:

   i)  identify the unit;

  ii)  establish the format (if any);

 iii)  position the file ready for the transfer (if required);

  iv)  transfer data between the file and the I/O list or namelist;

   v)  position the file following the transfer (if required);

  vi)  cause the IOSTAT and SIZE variables (if present) to become defined.

The order of transfer of namelist input is that in the input records.  Otherwise, the order is that of the I/O list or namelist.  Each input item is processed in turn, and may affect later subobjects and implied-DO indices.  All expressions within an I/O list item are determined at the beginning of processing of the item.  If an entity is specified more than once during execution of a namelist input statement, the later value overwrites the earlier value.  Any zero-sized array or zero-length character string is ignored.

When an input item is an array, no element of the array is permitted to affect the value of an expression within the item.  For example, the cases shown in Figure 43 are not permitted.  This prevents dependencies occurring within the item itself.

---

```
INTEGER J(10)
  :
READ *, J(J)                          ! Not permitted
READ *, J(J(1):J(10))                 ! Not permitted
```

---

Figure  43.

In the case of an internal file, an I/O item must not be in the file or associated with it. Nor may an input item contain or be associated with any portion of the established format.

Finally, a function reference must not appear in an expression anywhere in an I/O statement if it causes another I/O statement to be executed.

## 9.17 Summary

This chapter has begun the description of Fortran 90's extensive I/O facilities. It has covered the formatted I/O statements, and their associated format specifications, and then turned to unformatted I/O and direct-access files.

The syntax of the READ and WRITE statements has been intoduced gradually. The full syntax is

READ ( *control-list*) [*input-list*]

and

WRITE (*control-list*) [*output-list*]

where *control-list* contains one of more of the following: [UNIT=]*u*, [NML=]*nml-name*, REC=*i*, IOSTAT=*ios*, ERR=*error-label*, END=*end-label*, ADVANCE=*advance*, SIZE=*size*, EOR=*eor-label*. A *control-list* must include a unit specifier and must not include any specifier more than once. The IOSTAT and SIZE variables must not be associated with each other (for instance be identical), nor with any entity being transferred, nor with any *do-var* of an implied-DO list of the same statement. If either of these variables is an array element, the subscript value must not be affected by the data transfer, implied-DO processing, or the evaluation of any other specifier in the statement.

There are many detailed changes with respect to Fortran 77, often to support new features in other parts of the language, such as derived types. Other new features are NAMELIST, non-advancing I/O, the B, O, Z, EN and ES edit descriptors, and the generalization of the G edit descriptor.

## Exercises

**1.** Write suitable PRINT statements to print the name and contents of each of the following arrays:

(a) REAL GRID(10,10), 10 elements to a line (assuming the values are between 1.0 and 100.0);

(b) INTEGER LIST(50), the odd elements only;

(c) CHARACTER(LEN=10) TITLES(20), two elements to a line;

(d) REAL(10, 100) :: POWER(10), five elements to a line in engineering notation;

(e) LOGICAL FLAGS(10), on one line;

(f) COMPLEX PLANE(5), on one line.

**2.** Write statements to output the state of a game of tic-tac-toe (noughts-and-crosses) to a unit designated by the variable UNIT.

**3.** Write a program which reads an input record of up to 132 characters into an internal file and classifies it as a Fortran 90 comment line with no statement, an initial line without a statement label, an initial line with a statement label, a continuation line, or a line containing multiple statements.

**4.** Write separate list-directed input statements to fill each of the arrays of Exercise 1. For each statement write a sample first input record.

**5.** Write the function GET_CHAR, to read single characters from a formatted, sequential file, ignoring any record structure.

# 10. OPERATIONS ON EXTERNAL FILES

## 10.1 Introduction

So far we have discussed the topic of external files in a rather superficial way. In the examples of the various I/O statements in the previous chapter, an implicit assumption has always been made that the specified file was actually available, and that records could be written to it and read from it. For sequential files, the file control statements described in the next section further assume that it can be positioned. In fact, these assumptions are not necessarily valid. In order to define explicitly and to test the status of external files, three file status statements are provided: OPEN, CLOSE, and INQUIRE. Before beginning their description, however, two new definitions are required.

A computer system contains, among other components, a CPU and a storage system. Modern storage systems are usually based on some form of disc, which is used to store files for long or short periods of time. The execution of a computer program is, by comparison, a transient event. A file may exist for years, whereas programs run for only seconds or minutes. In Fortran terminology, a file is said to *exist* not in the sense we have just used, but in the restricted sense that it exists as a file *to which the program might have access*. In other words, if the program is prohibited from using the file because of a password protection system, or because some necessary action has not been taken in the job control language which is controlling the execution of the program, the file 'does not exist'.

A file which exists for a running program may be empty and may or may not be *connected* to that program. The file is connected if it is associated with a unit number known to the program. Such connection is usually made by executing an OPEN statement for the file, but many computer systems will *pre-connect* certain files which any program may be expected to use, such as terminal input and output. Thus we see that a file may exist but not be connected. It may also be connected but not exist. This can happen for a pre-connected new file. The file will only come into existence (be *created*) if some other action is taken on the file: executing an OPEN, WRITE, PRINT, or ENDFILE statement. A unit must not be connected to more than one file at once, and a file must not be connected to more than one unit at once.

There are a number of other points to note with respect to files:

- The set of allowed names for a file is processor dependent.

- Both sequential and direct access may be available for some files, but normally a file is limited to one or the other.

- A file never contains both formatted and unformatted records.

Finally, we note that no statement described in this chapter applies to internal files.

## 10.2  File positioning statements

When reading or writing an external file that is connected for sequential access, whether formatted or unformatted, it is sometimes necessary to perform other control functions on the file in addition to input and output. In particular, one may wish to alter the current position, which may be within a record, between records, ahead of the first record (at the *initial point*), or after the last record (at its *terminal point*). The following three statements are provided for these purposes.

## 10.2.1  BACKSPACE statement

It can happen in a program that a series of records is being written and that, for some reason, the last record written should be replaced by a new one, that is be overwritten. Similarly, when reading records, it may be necessary to reread the last record read, or to check-read a record which has just been written. For this purpose, Fortran provides the BACKSPACE statement, which has the syntax

BACKSPACE *u*

or

BACKSPACE ([UNIT=]*u* [,IOSTAT=*ios*] [,ERR=*error-label*])

where *u* is a scalar integer expression whose value is the unit number, and the other optional specifiers have the same meaning as for a READ statement. Again, keyword specifiers may be in any order, but the unit specifier must come first as a positional specifier.

The action of this statement is to position the file before the current record if it is positioned within a record, or before the preceding record if it is positioned between records. An attempt to backspace when already positioned at the beginning of a file results in no change in the file's position. If the file is positioned after an endfile record (Section 10.2.3), it becomes positioned before that record. It is not possible to backspace a file that does not exist, nor to backspace over a record written by a list-directed or namelist output statement (Sections 9.9 and 9.10). A series of BACKSPACE statements will backspace over the corresponding number of records. This statement is often very costly in computer resources and should be used as little as possible.

## 10.2.2  REWIND statement

In an analogous fashion to rereading, rewriting, or check-reading a record, a similar operation may be carried out on a complete file. For this purpose the REWIND statement,

REWIND *u*

or

REWIND ([UNIT=]*u* [,IOSTAT=*ios*] [,ERR=*error-label*])

may be used to reposition a file, whose unit number is specified by the scalar integer expression *u*. Again, keyword specifiers may be in any order, but the unit specifier must come first as a positional specifier. If the file is already at its beginning, there is no change in its position. The statement is permitted for a file that does not exist, and has no effect.

## 10.2.3  ENDFILE statement

The end of a file connected for sequential access is normally marked by a special record which is identified as such by the computer hardware, and computer systems ensure that all files written by a program are correctly terminated by such an *endfile record*. In doubtful situations, or when a subsequent program step will reread the file, it is possible to write an endfile record explicitly using the ENDFILE statement:

ENDFILE *u*

or

ENDFILE ([UNIT=] *u* [,IOSTAT=*ios*] [,ERR=*error-label*])

where *u*, once again, is a scalar integer expression specifying the unit number. Again, keyword specifiers may be in any order, but the unit specifier must come first as a positional specifier. The file is then positioned after the endfile record. This endfile record, if subsequently read by a program, must be handled using the END=*end-label* specifier of the READ statement, otherwise program execution will normally terminate. Prior to data transfer, a file must not be positioned after an endfile record, but it is possible to backspace or rewind across an endfile record, which allows further data transfer to occur.

An endfile record is written automatically whenever either a backspace or rewind operation follows a write operation as the next operation on the unit, or the file is closed by execution of a CLOSE statement (Section 10.4) or by normal program termination.

If the file may also be connected for direct access, only the records ahead of the endfile record are considered to have been written and only these may be read during a subsequent direct-access connection.

We note that if a file is connected to a unit, but does not exist for the program, it will be made to exist by executing an ENDFILE statement on the unit.

### 10.2.4  Data transfer statements

Execution of a data transfer statement (READ, WRITE, or PRINT) also affects the file position. If it is between records, it is moved to the start of the next record. Data transfer then takes place, which usually moves the position. No further movement occurs for non-advancing access. For advancing access, the position finally moves to follow the last record transferred.

## 10.3  OPEN statement

The OPEN statement is used to connect an external file to a unit, create a file that is preconnected, create a file and connect it to a unit, or change certain properties of a connection. The syntax is

OPEN ([UNIT=] *u, olist*)

where *u* is a scalar integer expression specifying the external file unit number, and *olist* is a list of optional specifiers. If the unit is specified with UNIT=, it may appear in *olist*. A specifier must not appear more than once. In the specifiers all entities are scalar and all characters are of default kind. In character expressions, any trailing blanks are ignored and, except for FILE=, any lower-case letters are converted to upper case. The specifiers are

**IOSTAT=** *ios*, where *ios* is a default integer variable which is set to zero if the statement is correctly executed, and to a positive value otherwise.

**ERR=**    *error-label*, where *error-label* is the label of a statement in the same scoping unit to which control will be transferred in the event of an error occurring during execution of the statement.

**FILE=**   *fln*, where *fln* is a character expression that provides the name of the file. If this specifier is omitted and the unit is not connected to a file, the STATUS= specifier must be specified with the value SCRATCH and the file connected to the unit will then depend on the computer system.   Whether the interpretation is case sensitive varies from system to system.

**STATUS=** *st*, where *st* is a character expression that provides the value OLD, NEW, REPLACE, SCRATCH, or UNKNOWN.   The FILE= specifier must be present if OLD, NEW, or REPLACE is specified; it must not be present if SCRATCH is specified.

If OLD is specified, the file must already exist; if NEW is specified, the file must not already exist, but will be brought into existence by the action of the OPEN statement.   The status of the file then becomes. OLD.

If REPLACE is specified and the file does not already exist, the file is created; if the file does exist, the file is deleted, and a new file is created with the same name.   In each case the status is changed to OLD.

If the value SCRATCH is specified, the file is created and becomes connected, but it cannot be kept after completion of the program or execution of a CLOSE statement (Section 10.4).

If UNKNOWN is specified, the status of the file is system dependent.   This is the default value of the specifier, if it is omitted.

**ACCESS=** *acc*, where *acc* is a character expression that provides the value SEQUENTIAL or DIRECT.   For a file which already exists, this value must be an allowed value.   If the file does not already exist, it will be brought into existence with the appropriate access method.   If this specifier is omitted, the value SEQUENTIAL will be assumed.

**FORM=** *fm*, where *fm* is a character expression that provides the value FORMATTED or UNFORMATTED, and determines whether the file is to be connected for formatted or unformatted I/O.   For a file which already exists, the value must be an allowed value.   If the file does not already exist, it will be brought into existence with an allowed set of forms that includes the specified form.   If this specifier is omitted, the default is FORMATTED for sequential access and UNFORMATTED for direct-access connection.

**RECL=** *rl*, where *rl* is an integer expression whose value must be positive. For a direct-access file, it specifies the length of the records, and is obligatory.   For a sequential file, it specifies the maximum length of a record, and is optional with a default value that is processor dependent.   For formatted files, the length is the number of characters for records that contain only default characters; for unformatted files it is system dependent but the INQUIRE statement (Section 10.5) may be used to find the length of an I/O list. In either case, for a file which already exists, the value specified must be allowed for that file. If the file does not already exist, the file will be brought into existence with an allowed set of record lengths that includes the specified value.

**BLANK=** *bl*, where *bl* is a character expression that provides the value NULL or ZERO. This connection must be for formatted I/O. This specifier sets the default for the interpretation of blanks in numeric input fields, as discussed in the description of the BN and BZ edit descriptors (Section 9.13.4). If the value is NULL, such blanks will be ignored (except that a completely blank field is interpreted as zero). If the value is ZERO, such blanks will be interpreted as zeros. If the specifier is omitted, the default is NULL.

**POSITION=** *pos*, where *pos* is a character expression that provides the value ASIS, REWIND, or APPEND. The access method must be sequential, and if the specifier is omitted the default value ASIS will be assumed. A new file is positioned at its initial point. If ASIS is specified and the file exists and is already connected, the file is opened without changing its position; if REWIND is specified the file is positioned at its initial point; if APPEND is specified and the file exists, it is positioned ahead of the endfile record if it has one (and otherwise at its initial point). For a file which exists but is not connected, the effect of the ASIS specifier on the file's position is unspecified.

**ACTION=** *act*, where *act* is a character expression that provides the value READ, WRITE, or READWRITE. If READ is specified, the WRITE, PRINT, and ENDFILE statements must not be used for this connection; if WRITE is specified, the READ statement must not be used; if READWRITE is specified, there is no restriction. If the specifier is omitted, the default value is processor dependent.

**DELIM=** *del* where *del* is a character expression that provides the value APOSTROPHE, QUOTE, or NONE. If APOSTROPHE or QUOTE is specified, the corresponding character will be used to delimit character constants written with list-directed or NAMELIST formatting, and it will be doubled where it appears within such a character constant; also non-default character values will be preceded by kind values. No delimiting character is used if NONE is specified, nor does any doubling take place. NONE is the default value if the specifier is omitted.  This specifier may appear only for formatted files.

**PAD=**  *pad*, where *pad* is a character expression that provides the value YES or NO.  If YES is specified, a formatted input record will be regarded as padded out with blanks whenever an input list and the associated format specify more data than appear in the record.  (If NO is specified, the length of the input record must not be less than that specified by the input list and the associated format, except in the presence of an ADVANCE='NO' specifier and either an EOR= or an IOSTAT= specification.)  The default value if the specifier is omitted is YES.  For non-default characters, the blank padding character is processor dependent.

An example of an OPEN statement is

```
OPEN (2, IOSTAT=IOS, ERR=99, FILE='CITIES',                    &
      STATUS='NEW', ACCESS='DIRECT', RECL=100)
```

which brings into existence a new, direct-access, unformatted file named CITIES, whose records have length 100.  The file is connected to unit number 2.  Failure to execute the statement correctly will cause control to be passed to the statement labelled 99, where the value of IOS may be tested.

The OPEN statements in a program are best collected together in one place, so that any changes which might have to be made to them when transporting the program from one system to another can be carried out without having to search for them.  Regardless of where they appear, the connection may be referenced in any program unit of the program.

The purpose of the OPEN statement is to connect a file to a unit.  If the unit is, however, already connected to a file then the action may be different.  If the FILE= specifier is omitted, the default is the name of the connected file.  If the file in question does not exist, but is pre-connected to the unit, then all the properties specified by the OPEN statement become part of the connection.  If the file is already connected to the unit, then of the existing attributes only the BLANK=, DELIM=, PAD=, ERR=, and IOSTAT= specifiers may have values different from those already in effect.  If the unit is already connected to another file, the effect of the OPEN statement includes the action of a prior CLOSE statement on the unit (without a STATUS= specifier, see below).

A file already connected to one unit must not be specified for connection to another unit.

In general, by repeated execution of the OPEN statement on the same unit, it is possible to process in sequence an arbitrarily high number of files, whether they exist or not, as long as the restrictions just noted are observed.

## 10.4 CLOSE statement

The purpose of the CLOSE statement is to disconnect a file from a unit. Its form is

CLOSE ([UNIT=]*u* [,IOSTAT=*ios*] [,ERR=*error-label*]    &
       [,STATUS=*st*])

where *u*, *ios*, and *error-label* have the same meanings as described above for the OPEN statement. Again, keyword specifiers may be in any order, but the unit specifier must come first as a positional specifier.

The function of the STATUS= specifier is to determine what will happen to the file once it is disconnected. The value of *st*, which is a scalar default character expression, may be either KEEP or DELETE, ignoring any trailing blanks and converting any lower-case letters to upper case. If the value is KEEP, a file that exists continues to exist after execution of the CLOSE statement, and may later be connected again to a unit. If the value is DELETE, the file no longer exists after execution of the statement. In either case, the unit is free to be connected again to a file. The CLOSE statement may appear anywhere in the program, and if executed for a non-existing or unconnected unit, acts as a 'do nothing' statement. The value KEEP must not be specified for files with the status SCRATCH.

If the STATUS= specifier is omitted, its default value is KEEP unless the file has status SCRATCH, in which case the default value is DELETE. On normal termination of execution, all connected units are closed, as if CLOSE statements with omitted STATUS= specifiers were executed.

An example of a CLOSE statement is

```
CLOSE (2, IOSTAT=IOS, ERR=99, STATUS='DELETE')
```

## 10.5 INQUIRE statement

The status of a file can be defined by the operating system prior to execution of the program, or by the program itself during execution, either by an OPEN statement or by some action on a pre-connected file which brings it into existence. At any time during the execution of a program it is possible to inquire about the status and attributes of a file using the INQUIRE statement. Using a variant of this statement, it is similarly possible to determine the status of a unit, for instance whether the unit number exists for that system (that is, whether it is an allowed unit number), whether the unit number has a file connected to it and, if so, which attributes that file has. Another variant permits an inquiry about the length of an output list when used to write an unformatted record.

Some of the attributes which may be determined by use of the INQUIRE statement are dependent on others. For instance, if a file is not connected to a unit, it is not meaningful to inquire about the form being used for that file. If this is nevertheless attempted, the relevant specifier is undefined.

The three variants are known as INQUIRE by file, INQUIRE by unit, and INQUIRE by output list. In the description of the INQUIRE statement which follows, the first two variants will be described together. Their forms are

INQUIRE ([UNIT=]*u, ilist*)

for INQUIRE by unit, where *u* is a scalar integer expression specifying an external unit, and

INQUIRE ( FILE=*fln, ilist*)

for INQUIRE by file, where *fln* is a scalar character expression whose value, ignoring any trailing blanks, provides the name of the file concerned. Whether the interpretation is case sensitive is system dependent. If the unit or file is specified by keyword, it may appear in *ilist*. A specifier must not occur more than once in the list of optional specifiers, *ilist*. All assignments occur following the usual rules, and all values of type character, apart from that for the NAME= specifier, are in upper case. The specifiers, in which all variables are scalar and of default kind, are

**IOSTAT=** *ios* and **ERR=** *error-label*, have the meanings described for them in the OPEN statement in Section 10.3. The IOSTAT= variable is the only one which is defined if an error condition occurs during the execution of the statement.

**EXIST=** *ex*, where *ex* is a logical variable. The value .TRUE. is assigned to *ex* if the file (or unit) exists, and .FALSE. otherwise.

**OPENED=** *open*, where *open* is a logical variable. The value .TRUE. is assigned to *open* if the file (or unit) is connected to a unit (or file), and .FALSE. otherwise.

**NUMBER=** *num*, where *num* is an integer variable that is assigned the value of the unit number connected to the file, or −1 if no unit is connected to the file.

**NAMED=** *nmd* and **NAME=** *nam*, where *nmd* is a logical variable that is assigned the value .TRUE. if the file has a name, and .FALSE. otherwise. If the file has a name, the character variable *nam* will be assigned the name. This value is not necessarily the same as that given in the FILE specifier, if used, but may be qualified in some way. However, in all cases it is a name which is valid for use in a subsequent OPEN statement, and so the INQUIRE can be used to determine the actual name of a file before connecting it. Whether the file name is case sensitive is system dependent.

**ACCESS=** *acc*, where *acc* is a character variable that is assigned the value SEQUENTIAL or DIRECT, depending on the access method for a file that is connected, and UNDEFINED if there is no connection.

**SEQUENTIAL=** *seq* and **DIRECT=** *dir*, where *seq* and *dir* are character variables that are assigned the value YES, NO, or UNKNOWN, depending on whether the file *may* be opened for sequential or direct access respectively, or whether this cannot be determined.

**FORM=** *frm*, where *frm* is a character variable that is assigned the value FORMATTED or UNFORMATTED, depending on the form for which the file is actually connected, and UNDEFINED if there is no connection.

**FORMATTED=** *fmt* and **UNFORMATTED=** *unf*, where *fmt* and *unf* are character variables that are assigned the value YES, NO, or UNKNOWN, depending on whether the file *may* be opened for formatted or unformatted access, respectively, or whether this cannot be determined.

**RECL=** *rec*, where *rec* is an integer variable that is assigned the value of the maximum record length allowed for the file. The length is the number of characters for formatted records containing only characters of default type, and system dependent otherwise. If there is no connection, *rec* becomes undefined.

**NEXTREC=** *nr*, where *nr* is an integer variable that is assigned the value of the number of the last record read or written, plus one. If no record has been yet read or written, it is assigned the value 1. If the file is not connected for direct access or if the position is indeterminate because of a previous error, *nr* becomes undefined.

**BLANK=** *bl*, where *bl* is a character variable that is assigned the value NULL or ZERO, depending on whether the blanks in numeric fields are by default to be interpreted as null fields or zeros, respectively, and UNDEFINED if there is either no connection, or if the connection is not for formatted I/O.

**POSITION=** *pos*, where *pos* is a character variable that is assigned the value REWIND, APPEND, or ASIS, as specified in the corresponding OPEN statement, if the file has not been repositioned since it was opened. If there is no connection, or if the file is connected for direct access, the value is UNDEFINED. If the file has been repositioned since the connection was established, the value is processor dependent (but must not be REWIND or APPEND unless that corresponds to the true position).

**ACTION=** *act*, where *act* is a character variable that is assigned the value READ, WRITE, or READWRITE, according to the connection. If there is no connection, the value assigned is UNDEFINED.

**READ=** *rd*, where *rd* is a character variable that is assigned the value YES, NO or UNKNOWN according to whether READ is allowed, not allowed, or is undetermined for the file.

**WRITE=** *wr*, where *wr* is a character variable that is assigned the value YES, NO or UNKNOWN according to whether WRITE is allowed, not allowed, or is undetermined for the file.

**READWRITE=** *rw*, where *rw* is a character variable that is assigned the value YES, NO or UNKNOWN according to whether READWRITE is allowed, not allowed, or is undetermined for the file.

**DELIM=** *del*, where *del* is a character variable that is assigned the value APOSTROPHE, QUOTE, or NONE, as specified by the corresponding OPEN statement (or by default). If there is no connection, or if the file is not connected for formatted I/O, the value assigned is UNDEFINED.

**PAD=** *pad*, where *pad* is a character variable that is assigned the value YES if so specified by the corresponding OPEN statement (or by default), and otherwise NO.

A variable that is a specifier in an INQUIRE statement or is associated with one must not appear in another specifier in the same statement.

The third variant of the INQUIRE statement, inquire by I/O list, has the form

INQUIRE (IOLENGTH=*length*) *olist*

where *length* is a scalar integer variable of default kind and is used to determine the length of an unformatted output list in processor-dependent units, and might be used to establish whether, for instance, an output list is too long for the record length given in the RECL= specifier of an OPEN statement, or be used as the value of the length to be supplied to a RECL= specifier, (see Figure 42 in Section 9.15).

An example of the INQUIRE statement, for the file opened as an example of the OPEN statement in Section 10.3, is

```
LOGICAL            EX, OP
CHARACTER (LEN=11) NAM, ACC, SEQ, FRM
INQUIRE (2, ERR=99, EXIST=EX, OPENED=OP, NAME=NAM, ACCESS=ACC,      &
    SEQUENTIAL=SEQ, FORM=FRM, RECL=IREC, NEXTREC=NR)
```

After successful execution of this statement, the variables provided will have been assigned the following values:

```
EX      .TRUE.
OP      .TRUE.
NAM     CITIESbbbbb
ACC     DIRECTbbbbb
SEQ     NObbbbbbbbb
FRM     UNFORMATTED
IREC    100
NR      1 (assuming no intervening read or write operations)
```

The three I/O status statements just described are perhaps the most indigestible of all Fortran 90 statements. They provide, however, a powerful and portable facility for the dynamic allocation and deallocation of files, completely under program control, which is far in advance of that found in any other programming language suitable for scientific applications.

## 10.6  Summary

This chapter has completed the description of the input/output features begun in the previous chapter, and together they provide a complete reference to all the facilities available. The new features described were INQUIRE by I/O list and the additional specifiers for the OPEN and INQUIRE statements: POSITION, ACTION, DELIM, PAD, READ, WRITE, and READWRITE. There have also been detailed changes to accommodate other features of the language.

## Exercises

**1.** A direct-access file is to contain a list of names and initials, to each of which there corresponds a telephone number. Write a program which opens a sequential file and a direct-access file, and copies the list from the sequential file to the direct-access file, closing it for use in another program.

Write a second program which reads an input record containing either a name or a telephone number (from a terminal if possible), and prints out the corresponding entry (or entries) in the direct-access file if present, and an error message otherwise. Remember that names are as diverse as Wu, O'Hara and Trevington-Smythe, and that it is insulting for a computer program to corrupt or abbreviate people's names. The format of the telephone numbers should correspond to your local numbers, but the actual format used should be readily modifiable to another.

# 11.  DEPRECATED FEATURES

## 11.1  Introduction

This chapter describes features which become redundant with the introduction of Fortran 90.  For most of them, their current heavy usage requires that they be retained for at least one further revision cycle before possibly becoming obsolescent, with final deletion a cycle later.  This is a decision that can be made only within the standardization process.  The remaining features are new ones that introduce a deliberate degree of redundancy into the language, and we have chosen to describe them along with the older redundant features.

Our deprecated features fall into three groups:

- those linked to storage association;

- those introduced into Fortran 90 because of strong public pressure, but for which there are better ways to achieve the same effects; and

- those which are made redundant by newer features.

Each description mentions how the feature concerned may be effectively replaced by a newer feature or features.

## 11.2  Storage association

*Storage units* are the fixed units of physical storage allocated to certain data. There is a storage unit called *numeric* for any nonpointer scalar of the default real, integer, and logical types, and a storage unit called *character* for any nonpointer scalar of type default character and character length 1.  Nonpointer scalars of type default complex or double precision real (Section 11.4.2) occupy two contiguous numeric storage units.  Nonpointer scalars of type default character length *len* occupy *len* contiguous character storage units.  As well as numeric and character storage units, there are a large number of *unspecified* storage units.  The standard makes no statement about the relative sizes of all these storage units and permits storage association to take place only between objects with the same category of storage unit.

Objects of derived type have no storage association, unless they have the SEQUENCE attribute:

```
TYPE STORAGE
   SEQUENCE
   INTEGER I
   REAL A(0:999)
END TYPE
```

Should any other derived types appear in such a definition, they too must have the SEQUENCE attribute. A derived type with the SEQUENCE attribute that does not have a pointer component at any level of component selection and whose ultimate components are of type default integer, default real, double precision real, default complex and default logical, has *numeric storage association*. Similarly, a derived type with the SEQUENCE attribute that does not have a pointer component at any level of component selection and whose ultimate components are of type default character has *character storage association*. Each other derived type with the SEQUENCE attribute and each non-default intrinsic type has its own unspecified storage unit.

A nonpointer array of intrinsic type or of derived type with the SEQUENCE attribute occupies a sequence of storage sequences, one for each element, in array element order. A nonpointer scalar of sequence type occupies a sequence of storage sequences corresponding to the sequence of its ultimate components. An object with the POINTER attribute has an unspecified storage unit, different from that of any nonpointer object and different for each combination of type, type parameters, and rank.

A sequence of storage sequences forms a storage sequence by concatenation.

A derived type with the SEQUENCE attribute may have private components:

```
TYPE STORAGE
   PRIVATE
   SEQUENCE
   INTEGER I
   REAL A(0:999)
END TYPE
```

The PRIVATE and SEQUENCE statements may be interchanged but must be the second and third statements of the type definition.

Two type definitions in different scoping units define the same data type if they have the same name, both have the SEQUENCE attribute, and they have components that are not PRIVATE and agree in order, name, and attributes. However, such a practice is prone to error and offers no advantage over having a single definition in a module and accessed by use association.

## 11.2.1  EQUIVALENCE statement

The EQUIVALENCE statement specifies that a given storage area may be shared by two or more objects. For instance

```
REAL AA, ANGLE, ALPHA, A(3)
EQUIVALENCE (AA, ANGLE), (ALPHA, A(1))
```

allows AA and ANGLE to be used interchangeably in the program text, as both names now refer to the same storage location. Similarly, ALPHA and A(1) may be used interchangeably.

It is possible to equivalence arrays together. In

```
REAL A(3,3), B(3,3), COL1(3), COL2(3), COL3(3)
EQUIVALENCE (COL1, A, B), (COL2, A(1,2)), (COL3, A(1,3))
```

the two arrays A and B are equivalenced, and the columns of A (and hence of B) are equivalenced to the arrays COL1, etc. We note in this example that more than two entities may be equivalenced together, even in a single declaration.

It is possible to equivalence variables of the same intrinsic type and kind type parameter or of the same derived type having the SEQUENCE attribute. It is also possible to equivalence variables of different types if both have numeric storage or both have character storage. Default character variables need not have the same length, as in

```
CHARACTER(LEN=4) A
CHARACTER(LEN=3) B(2)
EQUIVALENCE (A, B(1)(3:))
```

where the character variable A is equivalenced to the last four characters of the six characters of the character array B. An example for different types is

```
INTEGER I(100)
REAL X(100)
EQUIVALENCE (I, X)
```

where the arrays I and X are equivalenced. This might be used, for instance, to save storage space if I is used in one part of a program unit and X separately in another part. This is a highly dangerous practice, as considerable confusion can arise when one storage area contains variables of two or more data types, and program changes may be made very difficult if the two uses of the one area are to be kept distinct.

The general form of the statement is

EQUIVALENCE (*object, object-list*) [, (*object, object-list*)]...

where each *object* is a variable name, array element, or substring. An object must be a variable and must not be a dummy argument, a pointer, an object with a pointer component at any level of component selection, an allocatable array, an automatic object, a function, or a subobject of such an object. Each array subscript and character substring range must be an initialization expression. The interpretation of an array name is identical to that of its first element.

The objects in an equivalence set are said to be *storage associated*. Those of nonzero length share the same first storage unit. Those of zero length are associated with each other and with the first storage unit of those of nonzero length. EQUIVALENCE statements may cause other parts of the objects to be associated, but not such that different subobjects of the same object share storage. For example

```
REAL A, B(2)
EQUIVALENCE (A(1), B), (A(2), B)  ! Prohibited
```

is not permitted. Also, objects declared in different scoping units must not be equivalenced. For example

```
USE MY_MODULE, ONLY : XX
REAL BB
EQUIVALENCE(XX, BB)              ! Prohibited
```

is not permitted.

The various uses to which the EQUIVALENCE was put are replaced by automatic arrays, allocatable arrays, and pointers, (reuse of storage, Sections 6.4 and 6.7), pointers as aliases (storage mapping, Sections 6.12), and the TRANSFER function (mapping of one data type onto another, Section 8.9).

## 11.2.2  COMMON blocks

We have seen in Chapter 5 how two program units are able to communicate by passing variables, or values of expressions between them via argument lists or by using modules. It is also possible to define areas of storage known as COMMON blocks. Each has a storage sequence and may be either named or unnamed, as shown by the simplified syntax of the COMMON specification statement,

COMMON [/[ *cname*]/] *vlist*

in which *cname* is an optional name, and *vlist* is a list of variable names, each optionally followed by an array extent specification. An unnamed COMMON block is known as a *blank* COMMON block. Examples of each are

```
COMMON /HANDS/ NSHUFF, NPLAY, NHAND, CARDS(52)
```

and

```
COMMON // BUFFER(10000)
```

in which the named COMMON block HANDS defines a data area containing the quantities which might be required by the subroutines of a card playing program, and the blank COMMON defines a large data area which might be used by different routines as a buffer area.

The name of a COMMON block has global scope and must differ from that of any other global entity (external procedure, program unit, or COMMON block). It may, however, be the same as that of a local entity other than a named constant or intrinsic procedure.

No object in a COMMON block may have the PARAMETER attribute or be a dummy argument, an automatic object, an allocatable array, or a function. An array may have its extents declared either in the COMMON statement or in a type declaration or DIMENSION statement. If it is a nonpointer array, the extents must be declared explicitly and with constant specification expressions. An object of derived type must have the SEQUENCE attribute.

In order for a subroutine to access the variables in the data area, it is sufficient to insert the COMMON definition in each scoping unit which requires access to one or more of the entities in the list. In this fashion, the variables NSHUFF, NPLAY, NHAND and CARDS are made available to the those scoping units. No variable may appear more than once in all the COMMON blocks in a scoping unit.

Usually, a COMMON block contains identical variable names in all its appearances, but this is not necessary. In fact, the shared data area may be partitioned in quite different ways in different routines, using different variable names. They are said to be storage associated. It is thus possible for one subroutine to contain a declaration

```
COMMON /COORDS/ X, Y, Z, I(10)
```

and another to contain a declaration

```
COMMON /COORDS/ I, J, A(11)
```

This means that a reference to I(1) in the first routine is equivalent to a reference to A(2) in the second. This manner of coding is both untidy and dangerous, and every effort should be made to ensure that all declarations of a given COMMON block declaration are identical in every respect.

A further practice that is permitted but which we do not recommend is to mix different storage units in the same COMMON block. When this is done, each position in the storage sequence must always be occupied by a storage unit of the same category.

The total number of storage units must be the same in each occurrence of a named COMMON block, but blank COMMON is allowed to vary in size and the longest definition will apply for the complete program.

Yet another practice to be avoided is to use the full syntax of the COMMON statement,

COMMON [/[cname]/]vlist[ [,]/[cname]/vlist]...

which allows several COMMON blocks to be defined in one statement, and a single COMMON block to be declared in parts. A combined example is

```
COMMON /PTS/X,Y,Z /MATRIX/A(10,10),B(5,5) /PTS/I,J,K
```

which is equivalent to

```
COMMON /PTS/ X, Y, Z, I, J, K
COMMON /MATRIX/ A(10,10), B(5,5)
```

which is certainly a more understandable declaration of two shared data areas. The only need for the piece-wise declaration of one block is when the limit of 39 continuation lines is otherwise too low.

The COMMON statement may be combined with the EQUIVALENCE statement, as in the example

```
REAL A(10), B
EQUIVALENCE (A,B)
COMMON /CHANGE/ B
```

In this case, A is regarded as part of the COMMON block, and its length is extended appropriately. Such an equivalence must not cause data in two different COMMON blocks to become storage associated, it must not cause an extension of the COMMON block except at its tail, and two different objects or subobjects in the same COMMON block must not become storage associated.

A COMMON block may be declared in a module, in which case it must not be declared in a scoping unit accessing the module. Variable names in a COMMON block in a module may be declared to have the PRIVATE attribute.

An individual variable in a COMMON block may not be given the SAVE attribute, but the whole block may. If a COMMON block has the SAVE attribute in any scoping unit other than the main program, it must have the SAVE attribute in all such scoping units. The general form of the SAVE statement is

SAVE [[::] *saved-entity-list]*

where *saved-entity* is *variable-name* or *common-block-name*. A simple example is

```
SAVE /CHANGE/
```

Blank common always has the SAVE attribute.

Data in a COMMON block without the SAVE attribute become undefined on return from a subprogram unless the block is also declared in the main program or in another subprogram that is in execution.

Use of modules (Section 5.5) obviates the need for COMMON blocks.

## 11.2.3  BLOCK DATA

Nonpointer variables in named COMMON blocks may be initialized in DATA statements, but such statements must be collected into a special type of program unit, known as a BLOCK DATA program unit. It must have the form

BLOCK DATA [ *block-data-name]*
    [*specification-stmts*]
END [BLOCK DATA [*block-data- name*]]

where each *specification-stmt* is an IMPLICIT, PARAMETER, type declaration (including DOUBLE PRECISION), USE, INTRINSIC, POINTER, TARGET, COMMON, DIMENSION, EQUIVALENCE, DATA, or SAVE statement or derived-type definition. A type declaration statement must not specify the ALLOCATABLE, EXTERNAL, INTENT, OPTIONAL, PRIVATE, or PUBLIC attributes. An example is

```
BLOCK DATA
    COMMON /AXES/ I,J,K
    DATA I,J,K /1,2,3/
END
```

in which the variables in the COMMON block AXES are defined for use in any other scoping unit which accesses them.

It is possible to collect many COMMON blocks and their corresponding DATA statements together in one BLOCK DATA program unit. However, it may be a better practice to have several different BLOCK DATA program units, each containing COMMON blocks which have some logical association with one another. To allow for this eventuality, BLOCK DATA program units may be named in order to be able to distinguish them. A complete program may contain any number of BLOCK DATA program units, but only one of them may be unnamed. A COMMON block must not appear in more than one BLOCK DATA program unit. It is not possible to initialize blank COMMON.

The name of a BLOCK DATA program unit may appear in an EXTERNAL statement. When a processor is loading program units from a library, it may need such a statement in order to load the BLOCK DATA program unit.

Use of modules (Section 5.5) obviates the need for BLOCK DATA.

## 11.2.4  Shape and character length disagreement

In Fortran 77, it was often convenient, when passing an array, not to have to specify the size of the dummy array. For this case, the *assumed size* array declaration is available, where the last *extent* in the *extent-list* is

[*lower-bound*:] *

and the other extents (if any) must be declared explicitly. Such an array must not be a function result.

Since an assumed-size array has no extent in its last dimension, it does not have a shape and, therefore, must not be used as a whole array in an executable statement, except as an argument to a procedure that does not require its shape. However, if an array section is formed with an explicit upper bound in the last dimension, this has a shape and may be used as a whole array.

An object of one size or rank may be passed to an explicit-shape or assumed-size dummy argument array that is of another size or rank. If an array element is passed to an array, the actual argument is regarded as an array with elements that are formed from the parent array from the given array element onwards, in array element order. Figure 44 illustrates this. Here only the last 49 elements of A are available to SUB, as the first array element of A which is passed to SUB is A(52). Within SUB, this element is referenced as B(1).

```
REAL A(100)
:
CALL SUB (A(52), 49)
:
SUBROUTINE SUB(B,N)
:
REAL B(N)
:
```

Figure 44.

In the same example, it would also be perfectly legitimate for the declaration of B to be written as

```
REAL B(7, 7)
```

and for the last 49 elements of A to be addressed as though they were ordered as a 7×7 array. The converse is also true. An array dimensioned 10×10 in a calling subroutine may be dimensioned as a singly-dimensioned array of size 100 in the called subroutine. Within SUB, it is illegal to address B(50) in any way, as that would be beyond the declared length of A in the calling routine. In all cases, the association is by storage sequence, in array element order.

In the case of default character type, agreement of character length is not required. For a scalar dummy argument of character length *len*, the actual argument may have a greater character length and its leftmost *len* characters are associated with the dummy argument. For example, if CHASUB has a single dummy argument of character length 1,

```
CALL CHASUB(WORD(3:4))
```

is a valid CALL statement. For an array dummy argument, the restriction is on the total number of characters in the array. An array element or array element substring is regarded as a sequence of characters from its first character to the last character of the array. For an assumed-size array, the size is the number of characters in the sequence divided by the character length of the dummy argument.

Shape or character length disagreement of course cannot occur when the dummy argument is assumed-shape (by definition the shape is assumed from the actual argument). It can occur for explicit-shape and assumed-size arrays. Implementations are likely to receive explicit-shape and assumed-size arrays in contiguous storage, but permit any uniform spacing of the elements of an assumed-shape array. They will need to make a copy of any array argument that is not stored contiguously (for example, the section A(1:10:2)), unless the dummy argument is assumed shape. To avoid unnecessary copies of this kind, a scalar actual argument is permitted to be associated with an array only if the actual argument is an element of a named array that is explicit-shaped or assumed-sized, or a subobject of such an element.

When a procedure is invoked through a generic name, as a defined operation, or as a defined assignment, rank agreement between the actual and the dummy arguments is required. Note also that only a scalar dummy argument may be associated with a scalar actual argument.

Assumed-shape arrays (Section 6.3) supplant this feature.

## 11.2.5  ENTRY statement

A subprogram usually defines a single procedure, and the first statement to be executed is the first executable statement after the header statement. In some cases it is useful to be able to define several procedures in one subprogram, particularly when wishing to share access to some saved local variables or to a section of code. This is possible for external and module subprograms (but not for internal subprograms) by means of the ENTRY statement. This is a statement that has the form

```
ENTRY entry-name [([dummy-argument-list])       &
        [RESULT(result-name)]]
```

and may appear anywhere between the header line and CONTAINS (or END if it has no CONTAINS) statement of a subprogram, except within a construct. The ENTRY statement provides a procedure with an associated dummy argument list, exactly as does the SUBROUTINE or FUNCTION statement, and these arguments may be different from those given on the SUBROUTINE or FUNCTION statement. Execution commences with the first executable statement following the ENTRY statement.

In the case of a FUNCTION, each ENTRY defines another function, whose characteristics (that is, shape, type, type parameters, and whether a pointer) are given by specifications for the ENTRY name. If the characteristics are the same as for the main entry, a single variable is used for both results; otherwise, they must not be pointers, must be scalar, and must either both be of type default character with identical character length, or both be one of the default integer, default real, double precision real (11.4.2), or default complex types, and they are treated as equivalenced. The RESULT clause plays exactly the same rôle as for the main entry.

Each entry is regarded as defining another procedure, with its own name. The name of an entry has global scope and must differ from that of any other global entity (external procedure, program unit, or COMMON block). It must not be the name of a dummy argument of any of the procedures defined by the subprogram. In an interface block, there must be another body for each entry whose interface is wanted, using a SUBROUTINE or FUNCTION statement, rather than an ENTRY statement.

An ENTRY is called in exactly the same manner as a subroutine or function, depending on whether it appears in a subroutine subprogram or a function subprogram. An example is given in Figure 45 which shows a search function with two entry points. We note that LOOKU and LOOKS are synonymous within the function, so that it is immaterial which value is set before the return.

```
        FUNCTION LOOKU(LIST, MEMBER)
        INTEGER LOOKU, LIST(:), MEMBER, LOOKS
!
!       To locate MEMBER in an array LIST.
!       If LIST is unsorted, entry LOOKU is used,
!       if LIST is sorted, entry LOOKS is used.
!
!       List is unsorted
        DO LOOKU = 1, SIZE(LIST)
            IF LIST(LOOKU) .EQ. MEMBER) GO TO 9
        END DO
        GO TO 3
!
!       Entry for sorted list
        ENTRY LOOKS(LIST, MEMBER)
!
        DO LOOKU = 1, SIZE(LIST)
            IF (LIST(LOOKU) .GE. MEMBER) GO TO 2
        END DO
        GO TO 3
!
!       Is MEMBER at position LOOKU?
      2 IF (LIST(LOOKU) .EQ. MEMBER) GO TO 9
!
!       MEMBER not in LIST
      3 LOOKU = 0
!
      9 END
```

Figure 45.

None of the procedures defined by a subprogram is permitted to reference itself, unless the keyword RECURSIVE is present on the SUBROUTINE or FUNCTION statement. For a function, such a reference must be indirect unless there is a RESULT clause on the FUNCTION or ENTRY statement. If a procedure may be referenced directly in the subprogram that defines it, the interface is explicit in the subprogram.

The name of an ENTRY dummy argument that appears in an executable statement preceding the ENTRY statement in the subprogram must also appear in a FUNCTION, SUBROUTINE, or ENTRY statement that precedes the executable statement.

During the execution of one of the procedures defined by a subprogram, a reference to a dummy argument is permitted only if it is a dummy argument of the procedure referenced.

The ENTRY statement is made unnecessary by the use of modules (Section 5.5), with each procedure defined by an entry becoming a module procedure. Its presence has substantially complicated the standard itself because the reader has to remember that a subprogram may define several procedures.

## 11.3  New redundant features

### 11.3.1  INCLUDE line

It is sometimes useful to be able to include source text from somewhere else into the source stream presented to the compiler. This facility is possible using an INCLUDE line:

    INCLUDE *char-literal-constant*

where *char-literal-constant* must not have a kind parameter that is a named constant. This line is not a Fortran statement and must appear as a single source line where a statement may occur. It will be replaced by material in a processor-dependent way determined by the character string *char-literal-constant*. The included text may itself contain INCLUDE lines, which are similarly replaced. An INCLUDE line must not reference itself, directly or indirectly. When an INCLUDE line is resolved, the first included line must not be a continuation line and the last line must not be continued. An INCLUDE line may have a trailing comment, but may not be labelled nor, when expanded, may it contain incomplete statements.

There is no merit in using the INCLUDE facility as it is an uncontrolled mechanism; the same effect is better achieved with the USE statement (Section 5.5).

### 11.3.2  DO WHILE loop control

In Section 4.5, a form of the DO construct was described that may be written as

DO
    IF ( *scalar-logical-expr*) EXIT
    :
END DO

An alternative, but redundant, form of this is its representation using a DO WHILE statement:

DO [*label*] [,] WHILE (.NOT.*scalar-logical-expr*)

Potential optimization penalties that the use of the DO WHILE entails are fully described in Chapter 10 of *Optimizing Supercompilers for Supercomputers*, M. Wolfe (Pitman, 1989).

## 11.4  Old redundant features

### 11.4.1  Fixed source form

The fixed source form has been replaced by the free source form (Section 2.4). In the old form, each statement consists of one or more *lines* exactly 72 characters long,[15] and each line is divided into three *fields*. The first field consists of positions 1 to 5 and may contain a *statement label*. A Fortran statement may be written in the second fields of up to 20 consecutive lines. The first line of a multi-line statement is known as the *initial line*. and the succeeding lines as *continuation lines*.

A noncomment line is an initial line or a continuation line depending on whether there is a character, other than zero or blank, in position 6 of the line, which is the second field. The first field of a continuation line must be blank. The ampersand is not used for continuation.

The third field, from positions 7 to 72, is reserved for the Fortran statements themselves. Note that if a construct is named, the name must be placed here and not in the label field.

Except in a character context, blanks are insignificant.

The presence of an asterisk (*) or a character C in position 1 of a line indicates that the whole line is commentary. An exclamation mark indicates the start of commentary, except in position 6, where it indicates continuation.

---

[15] This limit is processor dependent if the line contains characters other than those of the default type.

Several statements separated by a semi-colon (;) may appear on one line. The semi-colon may not, in this case, be in column 6, where it would indicate continuation. Only the first of the statements on a line may be labelled.

A program unit END statement must not be continued, and no other statement may have an initial line that appears to be a program unit END statement.

A processor may restrict the appearance of its defined control characters, if any, in the fixed source form.

In applications where a high degree of compatibility between the old and the new source forms is required, for instance in code to be included into several programs which might exist in different forms, observance of the following rules can be of great help:

- confine statement labels to positions 1 to 5 and statements to positions 7 to 72;

- treat blanks as being significant;

- use only ! to indicate a comment (but not in position 6);

- for continued statements, place an ampersand in both position 73 of a continued line and position 6 of a continuing line.

## 11.4.2  Double precision real

Another *type* that may be used in a type declaration, FUNCTION, IMPLICIT, or component declaration statement is

DOUBLE PRECISION

which specifies double precision real. The precision is greater than that of default real.

Literal constants written with the exponent letter D are of type double precision real by default; no kind parameter may be specified if this exponent letter is used. Thus, 1D0 is of type double precision real. If DP is an integer named constant with the value KIND(1D0), DOUBLE PRECISION is synonymous with REAL(KIND=DP).

There is a D edit descriptor that was originally intended for double precision quantities but, in Fortran 90, it is identical to the E edit descriptor except that the output form may have a D instead of an E as its exponent letter. A double precision real literal constant, with exponent letter D, is acceptable on input whenever any other real literal constant is acceptable.

There are two elemental intrinsic functions which were not described in Chapter 8 because they have result of type double precision real:

**DBLE (A)** for A of type integer, real, or complex returns a double precision real value with as much precision of A (or the real part of A in the complex case) as possible.

**DPROD (X, Y)** returns the product X*Y for scalars X and Y of type default real as a double precision real result.

The double precision real data type has been replaced by the real type of kind KIND(0.D0).

## 11.4.3  Computed GO TO

A form of branch statement is the computed GO TO, which enables one path among many to be selected, depending on the value of a scalar integer expression. The general form is

GO TO ( *sl1, sl2, sl3,...*) [,] *intexpr*

where *sl1, sl2, sl3* etc. are labels of statements in the same scoping unit, and *intexpr* is any scalar integer expression. The same statement label may appear more than once. An example is

```
GO TO (6,10,20) I(K)**2+J
```

which references three statement labels. When the statement is executed, if the value of the integer expression is 1, the first branch will be taken, and control is transferred to the statement labelled 6. If the value is 2, the second branch will be taken, and so on. If the value is less than 1, or greater than 3, no branch will be taken, and the next statement following the GO TO will be executed.

This statement is replaced by the CASE construct (Section 4.4).

## 11.4.4  Character length specification *len

Alternatives to CHARACTER([LEN=]*len-value*) as a *type* in a type declaration, FUNCTION, IMPLICIT, or component definition statement are CHARACTER*(*len-value*)[,] and CHARACTER**len*[,], where *len* is an integer literal constant and the optional comma is permitted only when :: is absent. Note that this alternative form is provided only for default character.

### 11.4.5 Position of DATA statement

The DATA statement may be placed among the executable statements, but such placement offers no extra advantage and is rarely used and not recommended, since data initialization properly belongs with the specification statements.

### 11.4.6 Statement functions

It may happen that within a single program unit there are repeated occurrences of a computation which can be represented as a single statement. For instance, to calculate the parabolic function represented by

$$y = a + bx + cx^2$$

for different values of $x$, but with the same coefficients, there may be references to

```
Y1 = 1. + X1*(2. + 3.*X1)
:
Y2 = 1. + X2*(2. + 3.*X2)
:
```

etc. In Fortran 77, it was more convenient to invoke a so-called *statement function* (now better coded as an internal subroutine, Section 5.6), which must appear after any IMPLICIT and other relevant specification statements and before the executable statements. The example above would become

```
PARAB(X) = 1. + X*(2. + 3.*X)
:
Y1 = PARAB(X1)
:
Y2 = PARAB(X2)
```

Here, X is a dummy argument, which is used in the definition of the statement function. The variables X1 and X2 are actual arguments to the function.

The general form is

*function-name*([ *dummy-argument-list*]) = *scalar-expr*

where the *function-name* and each *dummy-argument* must be specified, explicitly or implicitly, to be scalar data objects. The *scalar-expr* must be composed of constants, references to scalar variables and array elements, references to functions, and intrinsic operations. A named constant that is referenced or an array of which an element is referenced must be declared earlier in the scoping unit. A scalar variable referenced may be a dummy argument of the statement function or a variable local to the scoping unit. A dummy argument of the host procedure must not be referenced unless it is a dummy argument of the main entry or of an ENTRY that precedes the statement function. If any entity is implicitly typed, a subsequent type declaration must confirm the type and type parameters. The dummy arguments are scalar and have a scope of the statement function statement only.

A statement function always has an implicit interface and may not be supplied as a procedure argument. It may appear within an internal procedure, and may reference other statement functions appearing before it in the same scoping unit, but not any appearing after. A function reference in the expression must not redefine a dummy argument.

Note that statement functions are irregular in that use and host association are not available.

## 11.4.7 Specific names of intrinsic procedures

There are a number of intrinsic functions that may have arguments that are all of one type and type parameters or all of another. For instance, we may write

```
A = SQRT(B)
```

and the appropriate square-root function will be invoked, depending on whether the variable B is real or complex and on its type parameters. In this case, the name SQRT is known as a *generic name*, meaning that the appropriate function is supplied, depending on the type and type parameters of the actual arguments of the function, and that a single name may be used for what are, probably, different specific functions.

Some of the specific functions have *specific names* and are specified by the standard. They are listed in Table 6 and Table 7. In the Tables, 'Character' stands for default character, 'Integer' stands for default integer, 'Real' stands for default real, 'Double' stands for double precision real, and 'Complex' stands for default complex. Those in Table 7 may be passed as actual arguments to a subprogram, provided they are specified in an INTRINSIC statement (Section 8.1.3).

All the procedures that we described in Chapter 8 are regarded as generic, even where there is only one version.

Table 6. Specific intrinsic functions that may not be used as actual arguments.

| Description | Generic Form | Specific Name | Argument Type | Function Type |
|---|---|---|---|---|
| Conversion to Integer | INT(A) | INT | Real | Integer |
|  |  | IFIX | Real | Integer |
|  |  | IDINT | Double | Integer |
| Conversion to Real | REAL(A) | REAL | Integer | Real |
|  |  | FLOAT | Integer | Real |
|  |  | SNGL | Double | Real |
| max(A1,A2,...) | MAX(A1,A2,...) | MAX0 | Integer | Integer |
|  |  | AMAX1 | Real | Real |
|  |  | DMAX1 | Double | Double |
|  |  | AMAX0 | Integer | Real |
|  |  | MAX1 | Real | Integer |
| min(A1,A2,...) | MIN(A1,A2,...) | MIN0 | Integer | Integer |
|  |  | AMIN1 | Real | Real |
|  |  | DMIN1 | Double | Double |
|  |  | AMIN0 | Integer | Real |
|  |  | MIN1 | Real | Integer |

Table 7 (Page 1 of 2). Specific intrinsic functions that may be used as actual arguments.

| Description | Generic Form | Specific Name | Argument Type | Function Type |
|---|---|---|---|---|
| Absolute value of A times sign of B | SIGN(A,B) | ISIGN | Integer | Integer |
| | | SIGN | Real | Real |
| | | DSIGN | Double | Double |
| max(X–Y,0) | DIM(X,Y) | IDIM | Integer | Integer |
| | | DIM | Real | Real |
| | | DDIM | Double | Double |
| X*Y | | DPROD(X,Y) | Real | Double |
| Truncation | AINT(A) | AINT | Real | Real |
| | | DINT | Double | Double |
| Nearest whole number | ANINT(A) | ANINT | Real | Real |
| | | DNINT | Double | Double |
| Nearest Integer | NINT(A) | NINT | Real | Integer |
| | | IDNINT | Double | Integer |
| Absolute value | ABS(A) | IABS | Integer | Integer |
| | | ABS | Real | Real |
| | | DABS | Double | Double |
| | | CABS | Complex | Real |
| Remainder modulo P | MOD(A,P) | MOD | Integer | Integer |
| | | AMOD | Real | Real |
| | | DMOD | Double | Double |
| Square root | SQRT(X) | SQRT | Real | Real |
| | | DSQRT | Double | Double |
| | | CSQRT | Complex | Complex |
| Exponential | EXP(X) | EXP | Real | Real |
| | | DEXP | Double | Double |
| | | CEXP | Complex | Complex |
| Natural logarithm | LOG(X) | ALOG | Real | Real |
| | | DLOG | Double | Double |
| | | CLOG | Complex | Complex |
| Common logarithm | LOG10(X) | ALOG10 | Real | Real |
| | | DLOG10 | Double | Double |

| | | | | |
|---|---|---|---|---|
| Sine | SIN(X) | SIN | Real | Real |
| | | DSIN | Double | Double |
| | | CSIN | Complex | Complex |
| Cosine | COS(X) | COS | Real | Real |
| | | DCOS | Double | Double |
| | | CCOS | Complex | Complex |
| Tangent | TAN(X) | TAN | Real | Real |
| | | DTAN | Double | Double |
| Arcsine | ASIN(X) | ASIN | Real | Real |
| | | DASIN | Double | Double |
| Arccosine | ACOS(X) | ACOS | Real | Real |
| | | DACOS | Double | Double |
| Arctangent | ATAN(X) | ATAN | Real | Real |
| | | DATAN | Double | Double |
| | ATAN2(Y,X) | ATAN2 | Real | Real |
| | | DATAN2 | Double | Double |
| Hyperbolic Sine | SINH(X) | SINH | Real | Real |
| | | DSINH | Double | Double |
| Hyperbolic Cosine | COSH(X) | COSH | Real | Real |
| | | DCOSH | Double | Double |
| Hyperbolic Tangent | TANH(X) | TANH | Real | Real |
| | | DTANH | Double | Double |
| Imaginary part | AIMAG(Z) | AIMAG | Complex | Real |
| Complex conjugate | CONJG(Z) | CONJG | Complex | Real |
| Character length | LEN(S) | LEN | Character | Integer |
| Starting position | INDEX(S,T) | INDEX | Character | Integer |

## 11.4.8  Assumed character length of function results

A nonrecursive external function whose result is scalar, character, and non-pointer may have assumed character length as in the example

```
FUNCTION COPY(WORD)
   CHARACTER(LEN=*) COPY, WORD
   COPY = WORD
END
```

Such a function is not permitted to specify a defined operation. In a scoping unit that invokes such a function, there must be an explicit declaration of the length, as in the example

```
PROGRAM MAIN
   CHARACTER(LEN=10) COPY
   WRITE(*, *) COPY('This message will be truncated')
END
```

or such a declaration must be accessible by use or host association.

This facility is included only for compatibility with Fortran 77 and is completely at variance with the philosophy of Fortran 90 that the attributes of a function result depend only on the actual arguments of the invocation and on any data accessible by the function through host or use association.

This facility may be replaced by use of a subroutine whose arguments correspond to the function result and the function arguments.

# Appendix A. INTRINSIC PROCEDURES

| Name | Section | Description |
|------|---------|-------------|
| ABS (A) | 8.3.1 | Absolute value. |
| ACHAR (I) | 8.5.1 | Character in position I of ASCII collating sequence. |
| ACOS (X) | 8.4 | Arc cosine (inverse cosine) function. |
| ADJUSTL (STRING) | 8.5.3 | Adjust left, removing leading blanks and inserting trailing blanks. |
| ADJUSTR (STRING) | 8.5.3 | Adjust right, removing trailing blanks and inserting leading blanks. |
| AIMAG (Z) | 8.3.1 | Imaginary part of complex number. |
| AINT (A [,KIND]) | 8.3.1 | Truncate to a whole number. |
| ALL (MASK [,DIM]) | 8.11 | True if all elements are true. |
| ALLOCATED (ARRAY) | 8.12.1 | True if the array is allocated. |
| ANINT (A [,KIND]) | 8.3.1 | Nearest whole number. |
| ANY (MASK [,DIM]) | 8.11 | True if any element is true. |
| ASIN (X) | 8.4 | Arcsine (inverse sine) function. |
| ASSOCIATED (POINTER [,TARGET]) | 8.2 | True if pointer is associated with target. |
| ATAN (X) | 8.4 | Arctangent (inverse tangent) function. |
| ATAN2 (Y, X) | 8.4 | Argument of complex number (X, Y). |
| BIT_SIZE (I) | 8.8.1 | Maximum number of bits that may be held in an integer. |
| BTEST (I, POS) | 8.8.2 | True if bit I of integer POS has value 1. |
| CEILING (A) | 8.3.1 | Least integer greater than or equal to its argument. |
| CHAR (I [,KIND]) | 8.5.1 | Character in position I of the processor collating sequence. |

| | | |
|---|---|---|
| CMPLX (X [,Y] [,KIND]) | 8.3.1 | Convert to COMPLEX type. |
| CONJG (Z) | 8.3.2 | Conjugate of a complex number. |
| COS (X) | 8.4 | Cosine function. |
| COSH (X) | 8.4 | Hyperbolic cosine function. |
| COUNT (MASK [,DIM]) | 8.11 | Number of true elements. |
| CSHIFT (ARRAY, SHIFT, [,DIM]) | 8.13.5 | Perform circular shift. |
| CALL DATE_AND_TIME ([DATE] [,TIME] [,ZONE] [,VALUES]) | 8.15.1 | Real-time clock reading date and time. |
| DBLE (A) | 11.4.2 | Convert to double precision real. |
| DIGITS (X) | 8.7.2 | Number of significant digits in the model for X. |
| DIM (X, Y) | 8.3.2 | Max(X-Y, 0). |
| DOT_PRODUCT (VECTOR_A, VECTOR_B) | 8.10 | Dotproduct. |
| DPROD (X, Y) | 11.4.2 | Double precision real product of two default real scalars. |
| EOSHIFT (ARRAY, SHIFT [,BOUNDARY] [,DIM]) | 8.13.5 | Perform end-off shift. |
| EPSILON (X) | 8.7.2 | Number that is almost negligible compared with one in the model for numbers like X. |
| EXP (X) | 8.4 | Exponential function. |
| EXPONENT (X) | 8.7.3 | Exponent part of the model for X. |
| FLOOR (A) | 8.3.1 | Greatest integer less than or equal to its argument. |
| FRACTION (X) | 8.7.3 | Fractional part of the model for X. |
| HUGE (X) | 8.7.2 | Largest number in the model for numbers like X. |
| IACHAR (C) | 8.5.1 | Position of character C in ASCII collating sequence. |
| IAND (I, J) | 8.8.2 | Logical AND on the bits. |
| IBCLR (I, POS) | 8.8.2 | Clear bit POS to zero. |
| IBITS (I, POS, LEN) | 8.8.2 | Extract a sequence of bits. |
| IBSET (I, POS) | 8.8.2 | Set bit POS to one. |

| | | |
|---|---|---|
| ICHAR (C) | 8.5.1 | Position of character C in the processor collating sequence. |
| IEOR (I, J) | 8.8.2 | Exclusive OR on the bits. |
| INDEX (STRING, SUBSTRING [,BACK]) | 8.5.3 | Starting position of SUBSTRING within STRING. |
| INT (A [,KIND]) | 8.3.1 | Convert to integer type. |
| IOR (I, J) | 8.8.2 | Inclusive OR on the bits. |
| ISHFT (I, ISHFT) | 8.8.2 | Logical shift on the bits. |
| ISHFTC (I, ISHFT [,SIZE]) | 8.8.2 | Logical circular shift on a set of bits on the right. |
| KIND (X) | 8.2 | Kind type parameter value. |
| LBOUND (ARRAY [,DIM]) | 8.12.2 | Array lower bounds. |
| LEN (STRING) | 8.6.1 | Character length. |
| LENTRIM (STRING) | 8.5.3 | Length of STRING without trailing blanks. |
| LGE (STRING_A, STRING_B) | 8.5.2 | True if STRING_A equals or follows STRING_B in ASCII collating sequence. |
| LGT (STRING_A, STRING_B) | 8.5.2 | True if STRING_A follows STRING_B in ASCII collating sequence. |
| LLE (STRING_A, STRING_B) | 8.5.2 | True if STRING_A equals or precedes STRING_B in ASCII collating sequence. |
| LLT (STRING_A, STRING_B) | 8.5.2 | True if STRING_A precedes STRING_B in ASCII collating sequence. |
| LOG (X) | 8.4 | Natural (base $e$) logarithm function. |
| LOGICAL (L, [,KIND]) | 8.5.4 | Convert between kinds of logicals. |
| LOG10 (X) | 8.4 | Common (base 10) logarithm function. |
| MATMUL (MATRIX_A, MATRIX_B) | 8.10 | Matrix multiplication. |
| MAX (A1, A2 [,A3,...]) | 8.3.2 | Maximum value. |
| MAXEXPONENT (X) | 8.7.2 | Maximum exponent in the model for numbers like X. |
| MAXLOC (ARRAY [,MASK]) | 8.14 | Location of maximum array element. |

| MAXVAL (ARRAY [,DIM] [,MASK]) | 8.11 | Value of maximum array element. |
| MERGE (TSOURCE, FSOURCE, MASK) | 8.13.1 | TSOURCE when MASK is true and FSOURCE otherwise. |
| MIN (A1, A2 [,A3,...]) | 8.3.2 | Minimum value. |
| MINEXPONENT (X) | 8.7.2 | Minimum exponent in the model for numbers like X. |
| MINLOC (ARRAY [,MASK]) | 8.14 | Location of minimum array element. |
| MINVAL (ARRAY [,DIM] [,MASK]) | 8.11 | Value of minimum array element. |
| MOD (A, P) | 8.3.2 | Remainder modulo P, that is A-INT(A/P)*P. |
| MODULO (A, P) | 8.3.2 | A modulo P. |
| CALL MVBITS (FROM, FROMPOS, LEN, TO, POS) | 8.8.3 | Copy bits. |
| NEAREST (X, S) | 8.7.3 | Nearest different machine number in the direction given by the sign of S. |
| NINT (A [,KIND]) | 8.3.1 | Nearest integer. |
| NOT (I) | 8.8.2 | Logical complement of the bits. |
| PACK (ARRAY, MASK [,VECTOR]) | 8.13.2 | Pack elements corresponding to true elements of MASK into rank-one result. |
| PRECISION (X) | 8.7.2 | Decimal precision in the model for X. |
| PRESENT (A) | 8.2 | True if optional argument is present. |
| PRODUCT (ARRAY [,DIM] [,MASK]) | 8.11 | Product of array elements. |
| RADIX (X) | 8.7.2 | Base of the model for numbers like X. |
| CALL RANDOM_NUMBER (HARVEST) | 8.15.2 | Random numbers in range $0 \leq x < 1$. |
| CALL RANDOM_SEED ([SIZE][,PUT] [,GET]) | 8.15.2 | Initialize or restart random number generator. |
| RANGE (X) | 8.7.2 | Decimal exponent range in the model for X. |
| REAL (A [,KIND]) | 8.3.1 | Convert to real type. |
| REPEAT (STRING, NCOPIES) | 8.6.2 | Concatenates NCOPIES of STRING. |

| RESHAPE (SOURCE, SHAPE [,PAD] [,ORDER]) | 8.13.3 | Reshape SOURCE to shape SHAPE. |
|---|---|---|
| RRSPACING (X) | 8.7.3 | Reciprocal of the relative spacing of model numbers near X. |
| SCALE (X, I) | 8.7.3 | $X \times b^I$, where $b$=RADIX(X). |
| SCAN (STRING, SET [,BACK]) | 8.5.3 | Index of left-most (right-most if BACK is true) character of STRING that belongs to SET; zero if none belong. |
| SELECTED_INT_KIND (R) | 8.7.4 | Kind of type parameter for specified exponent range. |
| SELECTED_REAL_KIND ([P] [,R]) | 8.7.4 | Kind of type parameter for specified precision and exponent range. |
| SET_EXPONENT (X, I) | 8.7.3 | Model number whose sign and fractional part are those of X and whose exponent part is I. |
| SHAPE (SOURCE) | 8.12.2 | Array (or scalar) shape. |
| SIGN (A, B) | 8.3.2 | Absolute value of A times sign of B. |
| SIN (X) | 8.4 | Sine function. |
| SINH (X) | 8.4 | Hyperbolic sine function. |
| SIZE (ARRAY [,DIM]) | 8.12.2 | Array size |
| SPACING (X) | 8.7.3 | Absolute spacing of model numbers near X. |
| SPREAD (SOURCE, DIM, NCOPIES) | 8.13.4 | NCOPIES copies of SOURCE forming an array of rank one greater. |
| SQRT (X) | 8.4 | Square root function. |
| SUM (ARRAY [,DIM] [,MASK]) | 8.11 | Sum of array elements. |
| CALL SYSTEM_CLOCK ([COUNT] [,COUNT_RATE] [,COUNT_MAX]) | 8.15.1 | Integer data from real-time clock. |
| TAN (X) | 8.4 | Tangent function. |
| TANH (X) | 8.4 | Hyperbolic tangent function. |
| TINY (X) | 8.7.2 | Smallest positive number in the model for numbers like X. |
| TRANSFER (SOURCE, MOLD [,SIZE]) | 8.9 | Same physical representation as SOURCE, but type of MOLD. |

| | | |
|---|---|---|
| TRANSPOSE (MATRIX) | 8.13.6 | Matrix transpose. |
| TRIM (STRING) | 8.6.2 | Remove trailing blanks from a single string. |
| UBOUND (ARRAY [,DIM]) | 8.12 | Array upper bounds. |
| UNPACK (VECTOR, MASK, FIELD) | 8.13.2 | Unpack elements of VECTOR corresponding to true elements of MASK. |
| VERIFY (STRING, SET [,BACK]) | 8.5.3 | Zero if all characters of STRING belong to SET or index of left-most (right-most if BACK true) that does not. |

# Appendix B.  FORTRAN 90 STATEMENTS

*Notes*:

- Obsolescent features (see Appendix C) have not been included in this list.
- Where no optional blank is indicated between two adjacent keywords, the blank is mandatory.

| Statement | Section |
| --- | --- |
| **NON-EXECUTABLE STATEMENTS** | |
| **Program Units and Subprograms** | |
| PROGRAM *program-name* | 5.2 |
| MODULE *module-name* | 5.5 |
| END[ ][MODULE [*module-name*]] | 5.5 |
| USE *module-name* [*,rename-list*] | 7.10 |
| USE *module-name*, ONLY:  [*only-list*] | 7.10 |
| PRIVATE [[::]*access-id-list*] | 7.6 & 7.11 |
| PUBLIC [[::] *access-id-list*] | 7.6 |
| EXTERNAL *external-name-list* | 5.11 |
| INTRINSIC *intrinsic-name-list* | 8.1.3 |
| [RECURSIVE] SUBROUTINE *subroutine-name* [([*dummy-argument-list*])] | 5.20 |
| [*prefix*] FUNCTION *function-name*([*dummy-argument-list*]) [RESULT(*result-name*)] | |
| where *prefix* is *type* [RECURSIVE] or RECURSIVE [*type*] | 5.20 |
| ENTRY *entry-name* [([*dummy-argument-list*]) [RESULT(*result-name*)]] | 11.2.5 |
| INTENT (*inout*) [::] *dummy-argument-name-list* | |
| where *inout* is IN, OUT, or IN[ ]OUT | 7.8 |
| OPTIONAL [::] *dummy-argument-name-list* | 7.8 |
| SAVE [ [::] *saved-entity-list*] | |
| where *saved-entity* is *variable-name* or /*common-block-name*/ | 7.9 |

TARGET [::] *object-name*[(*array-spec*)] [,*object-name*[(*array-spec*)] ]...    7.7

EQUIVALENCE (*object, object-list*) [, (*object, object-list*)]...    11.2.1

COMMON [/[*cname*]/]*vlist* [[,]/[*cname*]/ *vlist*]...    11.2.2

## EXECUTABLE STATEMENTS

### Assignment

*variable* = *expr*

where *variable* may be an array and may be a subobject    Chap. 3

*pointer* => *target*    3.12

IF (*scalar-logical-expr*) *action-stmt*    4.3.1

WHERE (*logical-array-expr*) *array-variable* = *array-expr*    6.8

### Program Units and Subprograms

CALL *subroutine-name* [([*actual-argument-list*])]    5.13

RETURN    5.8

END[ ][*unit*[*unit-name*]]    Chap. 5

where *unit* is PROGRAM, SUBROUTINE, or FUNCTION.

### Dynamic Storage Allocation

ALLOCATE (*allocation-list* [, STAT=*stat*] )    6.7.2

DEALLOCATE (*allocate-object-list* [, STAT=*stat*] )    6.7.3

NULLIFY (*pointer-object-list*)    6.7.4

### Control Constructs

[*do-name*:] DO [*label*] [,] *do-variable* = *scalar-integer-expr*,    4.5
*scalar-integer-expr* [,*scalar-integer-expr*]

[*do-name*:] DO [*label*] [,] WHILE(*scalar-logical-expr*)    11.3.2

CYCLE [*do-name*]    4.5

EXIT [*do-name*]    4.5

CONTINUE    4.5

END[ ]DO [*do-name*]    4.5

[*if-name*:] IF (*scalar-logical-expr*) THEN    4.3.2

ELSE[[ ]IF (*scalar-logical-expr*) THEN ] [*if-name*]    4.3.2

END[ ]IF [*if-name*]    4.3.2

[*select-name*:] SELECT[ ]CASE (*scalar-expr*)    4.4

| | |
|---|---|
| CASE (*case-value-list*) [*select-name*] | 4.4 |
| CASE DEFAULT [*select-name*] | 4.4 |
| END[ ]SELECT [*select-name*] | 4.4 |
| GO[ ]TO *label* | 4.2 |
| STOP [*access-code*] | 5.3 |
| WHERE (*logical-array-expr*) | 6.8 |
| ELSEWHERE | 6.8 |
| END[ ]WHERE | 6.8 |
| *function-name* ([*dummy-argument-list* ]) = *scalar-expr* | 11.4.6 |
| GO[ ]TO (*sl1, sl2, sl3,...*) [,] *intexpr* | 11.4.3 |

**Input-Output**

| | |
|---|---|
| READ (*control-list*) [*input-list*] | 9.17 |
| READ *format* [, *input-list*] | 9.7 |
| WRITE (*control-list*) [*output-list*] | 9.17 |
| PRINT *format* [, *output-list*] | 9.8 |
| REWIND *external-file-unit* | 10.2.2 |
| REWIND (*position-list*) | 10.2.2 |
| END[ ]FILE *external-file-unit* | 10.2.3 |
| END[ ]FILE (*position-list*) | 10.2.3 |
| BACKSPACE *external-file-unit* | 10.2.1 |
| BACKSPACE (*position-list*) | 10.2.1 |
| OPEN (*connect-list*) | 10.3 |
| CLOSE (*close-list*) | 10.4 |
| INQUIRE (*inquire-list*) | 10.5 |
| INQUIRE (IOLENGTH = *length*) *olist* | 10.5 |
| FORMAT ([*format-list*]) (this statement is actually non-executable). | 9.4 |

# Appendix C. OBSOLESCENT FEATURES

## C.1 Arithmetic IF

The arithmetic IF provides a three-way branching mechanism, depending on whether an arithmetic expression has a value which is less than, equal to, or greater than zero. It is replaced by the IF statement and construct (Section 4.3). Its general form is

IF (*expr*) *sl1, sl2, sl3*

where *expr* is any scalar expression of type integer or real, and *sl1, sl2,* and *sl3* are the labels of statements in the same scoping unit. If the result obtained by evaluating *expr* is negative then the branch to *sl1* is taken, if the result is zero the branch to *sl2*, and if the result is greater than zero the branch to *sl3*. An example is

```
      IF (P-Q) 1,2,3
   1  P = 0.
      GO TO 4
   2  P = 1.
      Q = 1.
      GO TO 4
   3  Q = 0.
   4  ...
```

in which a branch to 1, 2 or 3 is taken depending on the value of P–Q. The arithmetic IF may be used as a two-way branch when two of the labels are identical:

```
      IF (X-Y) 1,2,1
```

## C.2 DO-construct variations

The DO variable and the expressions that specify the limits and stride of a DO construct or an implied DO in an I/O statement may be of type default real or double precision real. We may therefore write a loop such as

```
      DO 1 A = 1, 15.7, 2.1
```

in which the real variable A will assume the initial value of 1.0 (note the conversion), and will subsequently have the values 3.1, 5.2, etc. up to 15.7.

There are, however, serious problems associated with DO-loops with real indices, and in order to understand them we recall how the number of iterations of a DO-loop is actually determined:

MAX(INT(( *expr2–expr1+expr3*)/*expr3*),0)

A consequence of this formula is rather insidious, and results from the application of the INT function. Consider the statement

```
DO 1 A = -0.3, -2.1, -0.3
```

which we would normally expect to result in seven iterations of the loop it controls. The number of iterations is obtained from the result of a computation whose intermediate value may not be 7.00000.. but 6.99999.., due to rounding errors. After applying the INT function we then have the integer 6 as the number of iterations. Whether or not this rounding error will occur for a given loop on a given computer is difficult to foresee, and this is the reason for avoiding the use of real DO-loop parameters. A similar problem can arise with a long loop: the repeated addition of the one to the other can lead to an unexpected loss of precision.

A DO-loop may be terminated on a labelled statement other than END DO or CONTINUE. Such a statement must be an executable statement other than GO TO, RETURN or an END statement of a subprogram, STOP or an END statement of a main program, EXIT, CYCLE, arithmetic IF, or assigned GO TO statement. Nested DO-loops may share the same labelled terminal statement, in which case all the usual rules for nested blocks hold, but a branch to the label must be from within the innermost loop. Thus we may write a matrix multiplication as

```
      A(1:N, 1:N) = 0.
      DO 1 I = 1,N
        DO 1 J = 1,N
          DO 1 L = 1,N
            A(I,J) = A(I,J)+B(I,L)*C(L,J)
  1   CONTINUE
```

Execution of a CYCLE statement restarts the loop without execution of the terminal statement. This form of DO-loop offers no additional functionality and considerable scope for unexpected mistakes.

## C.3  Assigned GO TO and assigned formats

Another form of branch statement is actually written in two parts, an ASSIGN statement and an assigned GO TO statement. The form is

ASSIGN *sl1* TO *intvar*

:

ASSIGN *sl2* TO *intvar*

:

GO TO *intvar* [[,](*sl1, sl2,...* )]

where *sl1, sl2* etc. are labels of statements in the same scoping unit, and *intvar* a scalar default integer variable. When an ASSIGN statement is executed, *intvar* acquires a representation of a statement label. Different labels may be assigned in different parts of the scoping unit. When the assigned GO TO is executed, then depending on the value of *intvar*, the appropriate path is taken.  If the optional statement label list in the GO TO statement is present, it must contain the label in *intvar*, and permits a check that *intvar* has acquired an expected value during program execution.  A label may appear more than once in the list.

The assigned GO TO's main purpose is to control logic flow in a scoping unit having a number of paths which come together at one point at which some common code is executed, and from which a new branch is taken depending on the path taken before.  This is shown in Figure 46, where we see three paths joining before the GO TO, and three after.  For this type of application an internal subprogram (Section 5.6) is more appropriate.

```
      :
      X = Y+1.
      ASSIGN 4 TO JUMP
      GO TO 3
  4   :
  1   X = Y+2.
      ASSIGN 5 TO JUMP
      GO TO 3
  5   :
  2   X = Y+3.
      ASSIGN 6 TO JUMP
  6   :
  3   Z = X**2
      :
      GO TO JUMP (4,5,6)
      :
```

Figure 46.

A default integer variable to which a statement label has been assigned in an ASSIGN statement may also be used to specify a FORMAT statement:

```
      ASSIGN 10 TO KEY
      :
      PRINT KEY, Q
  10  FORMAT(F10.3)
```

The label must be that of a FORMAT statement in the same scoping unit as the I/O statement.

This use of the ASSIGN statement is replaced by character expressions to define format specifiers (Section 9.4).

## C.4  Branching to an END IF statement

It is permissible to branch to an END IF statement from outside the construct that it terminates.  A branch to the following statement is a replacement for this practice.

## C.5 Alternate RETURN

When calling certain types of subroutines, it is possible that specific exceptional conditions will arise, which should cause a break in the normal control flow. It is possible to anticipate such conditions, and to code different flow paths following a subroutine call, depending on whether the called subroutine has terminated normally, or has detected an exceptional or abnormal condition. This is achieved using the alternate RETURN facility which uses the argument list in the following manner. Let us suppose that a subroutine DEAL receives in an argument list the number of cards in a shuffled deck, the number of players and the number of cards to be dealt to each hand. In the interests of generality, it would be a reasonable precaution for the first executable statement of DEAL to be a check that there is at least one player and that there are, in fact, enough cards to satisfy each player's requirement. If there are no players or insufficient cards, it can signal this to the main program which should then take the appropriate action. This may be written in outline as

```
      CALL DEAL(NSHUFF, NPLAY, NHAND, CARDS, *2, *3)
      CALL PLAY
      :
 2    ........   ! Handle no-player case
      :
 3    ........   ! Handle insufficient-cards case
      :
```

If the cards can be dealt, normal control is returned, and the call to PLAY executed. If an exception occurs, control is passed to the statement labelled 2 or 3, at which point some action must be taken —- to stop the game or shuffle more cards. The relevant statement label is defined by placing the statement label preceded by an asterisk as an actual argument in the argument list. It must be a label of an executable statement of the same scoping unit. Any number of such alternate returns may be specified, and they may appear in any position in the argument list. Since, however, they are normally used to handle exceptions, they are best placed at the end of the list.

   In the called subroutine, the corresponding dummy arguments are asterisks and the alternate RETURN is taken by executing a statement of the form

      RETURN *intexpr*

where *intexpr* is any scalar integer expression. The value of this expression at execution time defines an index to the alternate RETURN to be taken, according to its position in the argument list. If *intexpr* evaluates to 2, the second alternate RETURN will be taken. If *intexpr* evaluates to a value which is less than 1, or greater than the number of alternate RETURNs in the argument list, a normal RETURN will be taken. Thus, in DEAL, we may write simply

```
      SUBROUTINE DEAL(NSHUFF, NPLAY, NHAND, CARDS, *, *)
      :
      IF (NPLAY.LE.0) RETURN 1
      IF (NSHUFF .LT. NPLAY*NHAND) RETURN 2
```

This feature is also available for subroutines defined by ENTRY statements. It is not available for functions.

This feature is replaced by use of an integer argument holding a return code used in a computed GO TO following the CALL statement. A following CASE construct is now an even better alternative.

## C.6  PAUSE statement

At certain points in the execution of a program it might be useful to pause, in order to allow some possible external intervention in the running conditions to be made, for instance for an operator to activate a peripheral device required by the program. This can be achieved by executing a PAUSE statement which may also contain a default character constant, or a string of up to five digits, for example

```
PAUSE 'PLEASE MOUNT THE NEXT DISC PACK'
PAUSE 1234
```

Execution is resumed by some form of external command, for instance one given by an operator. The effect may be achieved with a READ statement that awaits data.

## C.7  H edit descriptor

The H edit descriptor provides an alternative character string edit descriptor: the output string is preceded by an $n$H edit descriptor, where $n$ is the number of default characters in the following string (blanks being significant):

```
100  FORMAT(23HI MUST COUNT CHARACTERS)
```

The value $n$ must be an integer literal constant without a kind parameter. If the Hollerith string occurs within a character constant delimited by apostrophes and contains an apostrophe, the apostrophe must be represented by two apostrophes, but counts as only one in the $n$H character count, as in the example

```
PRINT '(7H DON''T, A)', CAUTION
```

and similarly for quotes.

The H edit descriptor provides the same functionality as the character string edit descriptor but is prone to error as it is easy to miscount the number of characters in the string.

# Appendix D.  FORTRAN TERMS

The following is a list of the principal technical terms used in this book and their definitions.  To facilitate reference to the draft standard, we have kept closely to the meanings used there.  Where the definition uses a term that is itself defined in this glossary, the first occurrence of the term is printed in italics.  Some terms used in Fortran 77 have a different meaning here and we draw the reader's attention to each such term by marking it with a bold asterisk **.  We make no reference to obsolescent features (Appendix C) in this Appendix.

**Actual argument**.  An *expression*, a *variable*, or a *procedure* that is specified in a procedure *reference*.

**Allocatable array**.  A *named array* having the ALLOCATABLE *attribute*.  Only when it has space allocated for it does it have a *shape* and may it be *referenced* or *defined*.

**Argument**.  An *actual argument* or a *dummy argument*.

**Argument association**.  The relationship between an *actual argument* and a *dummy argument* during the execution of a *procedure reference*.

**Argument keyword**.  A *dummy argument name*.  It may used in a *procedure reference* ahead of the equals symbol provided the procedure has an *explicit interface*.

**Array** **.  A set of *scalar data*, all of the same *type* and *type parameters*, whose individual elements are arranged in a rectangular pattern.  It may be a *named* array, the *target* of an *array pointer*, an *array section*, a *structure component*, a *function* value, or an *expression*.  Its *rank* is at least one.  [In Fortran 77, arrays were always named and always variables.]

**Array element**.  One of the *scalar data* that make up an *array*.

**Array pointer**.  A *pointer* to an *array*.

**Array section**.  A *subobject* of an *array* consisting of a set of *array elements* or *substrings* of a set of array elements.  The set is specified by *subscripts*, *subscript triplets*, and *vector subscripts*.

**Array-valued**.  Having the property of being an *array*.

**Assignment statement**.  A *statement* of the form '*variable = expression*'.

**Assignment token**.  The *lexical token* = used in an *assignment statement*.

**Association**.  *Name association*, *pointer association*, or *storage association*.

**Assumed-size array.** A *dummy array* whose *size* is assumed from the associated *actual argument*. Its last upper bound is specified by an asterisk.

**Attribute.** A property of a *data object* that may be specified in a *type declaration statement*.

**Belong.** If an EXIT or a CYCLE *statement* contains a *construct name*, the statement **belongs** to the DO construct using that name. Otherwise, it **belongs** to the innermost DO construct in which it appears.

**Block.** A sequence of *executable constructs* embedded in another executable construct, bounded by *statements* that are particular to the construct, and treated as an integral unit.

**Block data program unit.** A *program unit* that provides initial values for *data objects* in *named common blocks*.

**Bounds.** For a *named array*, the limits within which the values of the *subscripts* of its *array elements* must lie.

**Character.** A letter, digit, or other symbol.

**Characteristics.**

 i) Of a *procedure*, its classification as a *function* or *subroutine*, the characteristics of its *dummy arguments*, and the characteristics of its *function result* if it is a function.

 ii) Of a *dummy argument*, whether it is a *data object*, is a *procedure*, or has the OPTIONAL *attribute*.

 iii) Of a *data object*, its *type, type parameters, shape*, the exact dependence of an array bound or the character length on other entities, *intent*, whether it is optional, whether it is a *pointer* or a *target*, and whether the *shape, size*, or *character length* is assumed.

 iv) Of a *dummy procedure*, whether the interface is explicit, its characteristics as a procedure if the interface is explicit, and whether it is optional.

 v) Of a *function result*, its type, type parameters, whether it is a pointer, rank if it is a pointer, shape if it is not a pointer, the exact dependence of an array bound or the character length on other entities, and whether the character length is assumed.

**Character string.** A sequence of *characters* numbered from left to right 1, 2, 3,...

**Character storage unit.** The unit of storage for holding a *scalar* of *type* default character and character length one that is not a *pointer*.

**Collating sequence.** An ordering of all the different *characters* of a particular *kind type parameter*.

**Common block.** A block of physical storage that may be accessed by any of the *scoping units* in an *executable program*.

**Component.** A constituent of a *derived type*.

**Conformable.** Two *arrays* are said to be **conformable** if they have the same *shape*. A *scalar* is **conformable** with any array.

**Conformance.** An *executable program* conforms to the standard if it uses only those forms and relationships described therein and if the executable program has an interpretation according to the standard. A *program unit* conforms to the standard if it can be included in an executable program in a manner that allows the executable program to be standard conforming. A *processor* conforms to the standard if it executes standard-conforming programs in a manner that fulfills the interpretations prescribed in the standard.

**Connected.**

i) For an *external unit*, the property of referring to an *external file*.

ii) For an *external file*, the property of having an *external unit* that refers to it.

**Constant \*.** A *data object* whose value must not change during execution of an *executable program*. It may be a *named constant* or a *literal constant*. [In Fortran 77, constants were always literal constants.]

**Constant expression.** An *expression* satisfying rules that ensure that its value does not vary during program execution.

**Construct.** A sequence of *statements* starting with a CASE, DO, IF, or WHERE statement and ending with the corresponding terminal statement.

**Data.** Plural of *datum*.

**Data entity.** An *entity* that has or may have a data value. It may be a *constant*, a *variable*, an *expression*, or a *function result*.

**Data object.** A *datum* of *intrinsic* or *derived type* or an *array* of such *data*. It may be a *literal constant*, a *named* data object, a *target* of a *pointer*, or it may be a *subobject*.

**Data type.** A *named* category of *data* that is characterized by a set of values, together with a way to denote these values and a collection of *operations* that interpret and manipulate the values. For an *intrinsic* data type, the set of data values depends on the values of the *type parameters*.

**Datum.** A single quantity that may have any of the set of values specified for its *data type*.

**Definable.** A *variable* is **definable** if its value may be changed by the appearance of its *name* or *designator* on the left of an *assignment statement*. A *allocatable array* that has not been allocated is an example of a *data object* that is not definable. An example of a *subobject* that is not definable is C(I) when C is an *array* that is a *constant* and I is an integer variable.

**Defined.** For a *data object*, the property of having or being given a valid value.

**Defined assignment statement.** An *assignment statement* that is not an *intrinsic* assignment statement and is defined by a *subroutine subprogram* and an *interface block*.

**Defined operation.** An *operation* that is not an *intrinsic* operation and is defined by a *function subprogram* and an *interface block*.

**Deleted feature.** A feature in Fortran 77 that is considered to have been redundant and largely unused. No features in Fortran 77 have been deleted from the standard. Note that a feature designated as an *obsolescent feature* in the standard may become a deleted feature in the next revision.

**Derived type.** A *type* whose *data* have *components* each of which is either of *intrinsic* type or of another derived type.

**Designator.** See *subobject designator*.

**Disassociated.** A *pointer* is **disassociated** following execution of a DEALLOCATE or NULLIFY *statement*.

**Dummy argument.** An entity whose *name* appears in the parenthesized list following the *procedure* name in a FUNCTION *statement*, a SUBROUTINE statement, an ENTRY statement, or a *statement function* statement.

**Dummy array.** A *dummy argument* that is an *array*.

**Dummy pointer.** A *dummy argument* that is a *pointer*.

**Dummy procedure.** A *dummy argument* that is specified or *referenced* as a *procedure*.

**Elemental.** An adjective applied to an *intrinsic operation, procedure*, or *assignment statement* that is applied independently to the elements of an *array* or corresponding elements of a set of *conformable* arrays and *scalars*.

**Entity.** The term used for any of the following: a *program unit*, a *procedure*, an *operator*, an *interface block*, a *common block*, an *external unit*, a *statement function*, a *type*, a *named variable*, an *expression*, a *component* of a *structure*, a *named constant*, a *statement label*, a *construct*, or a namelist group.

**Executable construct.** A CASE, DO, IF, or WHERE *construct*.

**Executable program.** A set of *program units* that includes exactly one *main program*.

**Executable statement.** An instruction to perform or control one or more computational actions.

**Explicit interface.** For a *procedure referenced* in a *scoping unit*, the property of being a *module procedure*, an *intrinsic procedure*, an *external procedure* that has an *interface block* or is defined by the scoping unit and is recursive, or a *dummy procedure* that has an interface block.

**Explicit-shape array.** A *named array* that is declared with *explicit bounds*.

**Expression.** A sequence of *operands*, *operators*, and parentheses. It may be a *variable*, a *constant*, a *function reference*, or may represent a computation.

**Extent.** The size of one dimension of an *array*.

**External file.** A sequence of *records* that exists in a medium external to the *executable program*.

**External procedure.** A *procedure* that is defined by an *external subprogram* or by a means other than Fortran.

**External subprogram \*.** A *subprogram* that is not contained in a *main program*, *module*, or another subprogram. [In Fortran 77, a *block data program unit* is called a subprogram.]

**External unit.** A mechanism that is used to refer to an *external file*. It is identified by a nonnegative integer.

**File.** An *internal file* or an *external file*.

**Function.** A *procedure* that is invoked in an *expression*.

**Function result.** The *data object* that returns the value of a *function*.

**Function subprogram.** A sequence of *statements* beginning with a FUNCTION statement that is not in an *interface block* and ending with the corresponding END statement.

**Generic identifier.** A *name*, *operator*, or *assignment token* specified in an INTERFACE *statement* to provide an alternative means of invoking any of the *procedures* in the *interface block*.

**Global entity.** An *entity* identified by a *lexical token* whose *scope* is an *executable program*. It may be a *program unit*, a *common block*, or an *external procedure*.

**Host.** A *main program* or *subprogram* that contains an *internal procedure* is called the **host** of the internal procedure. A *module* that contains a *module procedure* is called the **host** of the module procedure.

**Host association.** The process by which an *internal subprogram* or *derived type* definition accesses *entities* of its *host*.

**Implicit interface.** A *procedure referenced* in a *scoping unit* is said to have an **implicit interface** if the procedure does not have an *explict interface* there.

**Inquiry function.** An *intrinsic function* whose result depends on properties of the principal *argument* other than the value of the argument.

**Instance of a subprogram.** The copy of a *subprogram* that is created when a *procedure* defined by the subprogram is *invoked*.

**Intent.** Of a *dummy argument* that is a neither a *procedure* nor a *pointer*, whether it is intended to transfer data into the procedure, out of the procedure, or both.

**Interface block.** A sequence of *statements* beginning with a INTERFACE statement and ending with the corresponding END INTERFACE statement.

**Interface body.** A sequence of *statements* in an *interface block* beginning with a FUNCTION or SUBROUTINE statement and ending with the corresponding END statement.

**Interface of a procedure.** See *procedure interface*.

**Internal file.** A character *variable* that is used to transfer and convert *data* from internal storage to internal storage.

**Internal procedure.** A *procedure* that is defined by an *internal subprogram*.

**Internal subprogram.** A *subprogram* contained in a *main program* or another subprogram.

**Intrinsic.** An adjective applied to *types*, *operations*, *assignment statements*, and *procedures* that are defined in the standard and may be used in any *scoping unit* without further definition or specification.

**Invoke.**

   i) To call a *subroutine* by a CALL *statement* or by a *defined assignment statement*.

   ii) To call a *function* by a *reference* to it by *name* or *operator* during the evaluation of an *expression*.

**Keyword.** *Statement keyword* or *argument keyword*.

**Kind type parameter.** A parameter whose values label the available kinds of an *intrinsic type*.

**Label.** See *statement label*.

**Length of a character string.** The number of *characters* in the *character string*.

**Lexical token.** A sequence of one or more characters with an indivisible interpretation.

**Line.** A source-form *record* containing from 0 to 132 *characters*.

**Literal constant.** A *constant* without a *name*.

**Local entity.** An *entity* identified by a *lexical token* whose *scope* is a *scoping unit*.

**Main program.** A *program unit* that is not a *module, subprogram,* or *block data program unit.*

**Many-one array section.** An *array section* with a *vector subscript* having two or more elements with the same value.

**Module.** A *program unit* that contains or accesses definitions to be accessed by other program units.

**Module procedure.** A *procedure* that is defined by a *module subprogram.*

**Module subprogram.** A *subprogram* that is contained in a *module* but is not an *internal subprogram.*

**Name \*.** A *lexical token* consisting of a letter followed by up to 30 alphanumeric characters (letters, digits, and underscores). [In Fortran 77, this was called a symbolic name.]

**Name association.** *Argument association, use association,* or *host association.*

**Named.** Having a *name.*

**Named constant \*.** A *constant* that has a *name.* [In Fortran 77, this was called a symbolic constant.]

**Numeric storage unit.** The unit of storage for holding a *scalar* of *type* default real, default integer, or default logical that is not a *pointer.*

**Numeric type.** Integer, real, or complex *type.*

**Object.** *Data object.*

**Obsolescent feature.** A feature in Fortran 77 that is considered to have been redundant but that is still in frequent use.

**Operand.** An *expression* that precedes or succeeds an *operator*.

**Operation.** (7.1.2). A computation involving one or two *operands*.

**Operator.** A *lexical token* that specifies an *operation*.

**Pointer.** A *data object* that has the POINTER *attribute*. It may not be *referenced* or *defined* unless it is *pointer associated* with a *target*. If it is an *array*, it does not have a *shape* unless it is pointer associated.

**Pointer assignment.** The *pointer association* of a *pointer* with a *target* by the execution of a *pointer assignment statement* or the execution of an *assignment statement* for a *data object* of *derived type* having the pointer as a *subobject*.

**Pointer assignment statement.** A *statement* of the form '*pointer* => *target*'.

**Pointer associated.** The relationship between a *pointer* and a *target* following a *pointer assignment* or a valid execution of an ALLOCATE *statement*.

**Pointer association.** The process by which a *pointer* becomes *pointer associated* with a *target*.

**Present.** A *dummy argument* is **present** in an *instance* of a *subprogram* if it is *associated* with an *actual argument* and the actual argument is a dummy argument that is present in the invoking *procedure* or is not a dummy argument of the invoking procedure.

**Procedure.** A computation that may be *invoked* during program execution. It may be a *function* or a *subroutine*. It may be an *internal procedure*, an *external procedure*, a *module procedure*, a *dummy procedure*, or a *statement function*. A *subprogram* may define more than one procedure if it contains ENTRY *statements*.

**Procedure interface.** The *characteristics* of a *procedure*, the *name* of the procedure, the name of each *dummy argument*, and the *generic identifiers* (if any) by which it may be *referenced*.

**Processor.** The combination of a computing system and the mechanism by which *executable programs* are transformed for use on that computing system.

**Program.** See *executable program* and *main program*.

**Program unit.** The fundamental component of an *executable program*. A sequence of *statements* and comment lines. It may be a *main program*, a *module*, an *external subprogram*, or a *block data program unit*.

**Rank.** The number of dimensions of an *array*. Zero for a *scalar*.

**Record.** A sequence of values that is treated as a whole within a *file* .

**Reference.** The appearance of a *data object name* or *subobject designator* in a context requiring the value at that point during execution, or the appearance of a *procedure* name, its *operator* symbol, or a *defined assignment statement* in a context requiring execution of the procedure at that point. Note that neither the act of defining a *variable* nor the appearance of the name of a procedure as an *actual argument* is regarded as a reference.

**Scalar.**

i) A single *datum* that is not an *array*.

ii) Not having the property of being an *array*.

**Scope.** That part of an *executable program* within which a *lexical token* has a single interpretation. It may be an *executable program*, a *scoping unit*, a single *statement*, or a part of a statement.

**Scoping unit.** One of the following:

i) A *derived type* definition,

ii) An *interface body*, excluding any derived-type definitions and interface bodies contained within it, or

iii) A *program unit* or *subprogram*, excluding derived-type definitions, interface bodies, and subprograms contained within it.

**Section subscript.** A *subscript, subscript triplet*, or *vector subscript* in an *array section selector*.

**Selector.** A syntactic mechanism for designating

i) Part of a *data object*. It may designate a *substring*, an *array element*, an *array section*, or a *structure component*.

ii) The set of values for which a CASE *block* is executed.

**Shape.** For an *array*, the *rank* and *extents*. The shape may be represented by the rank-one array whose elements are the extents in each dimension.

**Size.** For an *array*, the total number of elements.

**Standard module.** A *module* standardized as a separate collateral standard.

**Statement.** A sequence of *lexical tokens*. It usually consists of a single line, but the ampersand symbol may be used to continue a statement from one line to another and the semicolon symbol may be used to separate statements within a line.

**Statement entity.** An *entity* identified by a *lexical token* whose *scope* is a single *statement* or part of a statement.

**Statement function.** A *procedure* specified by a single *statement* that is similar in form to an *assignment statement.*

**Statement keyword.** A word that is part of the syntax of a *statement* and that may be used to identify the statement.

**Statement label.** A *lexical token* consisting of up to five digits that precedes a *statement* and may be used to refer to the statement.

**Storage association.** The relationship between two *storage sequences* if a storage unit of one is the same as a storage unit of the other.

**Storage sequence.** A sequence of contiguous *storage units.*

**Storage unit.** A *character storage unit,* a *numeric storage unit,* or an *unspecified storage unit.*

**Stride.** The increment specified in a *subscript triplet.*

**Structure.** A *scalar data object* of *derived type.*

**Structure component.** The part of an *object* of *derived-type* corresponding to a *component* of its type.

**Subobject.** Of a *named data object* or *target* of a *pointer,* a portion that may be *referenced* or *defined* independently of other portions. It may be an *array element,* an *array section,* a *structure component,* or a *substring.*

**Subobject designator.** A *name,* followed by one or more *component selectors, array section* selectors, *array element* selectors, and *substring* selectors.

**Subprogram \*.** A *function subprogram* or a *subroutine subprogram.* [In Fortran 77, a *block data program unit* was called a subprogram.]

**Subroutine.** A *procedure* that is *invoked* by a CALL *statement* or by a *defined assignment statement.*

**Subroutine subprogram.** A sequence of *statements* from a SUBROUTINE statement that is not in an *interface block* to the corresponding END statement.

**Subscript \*.** One of the list of *scalar* integer *expressions* in an *array element selector.* [In Fortran 77, the whole list was called the subscript.]

**Subscript triplet.** An item in the list of an *array section selector* that contains a colon and specifies a regular sequence of integer values.

**Substring**. A contiguous portion of a *scalar character string*. Note that an *array section* can include a *substring selector*; the result is called an array section and not a substring.

**Target**. A *named data object* specified in a *type declaration statement* containing the TARGET *attribute*, a data object created by an ALLOCATE statement for a *pointer*, or a *subobject* of such an object.

**Transformational function**. An *intrinsic function* that is neither an *elemental* function nor an *inquiry function*. It usually has *array arguments* and an array result whose elements have values that depend on the values of many of the elements of the arguments.

**Type**. *Data type*.

**Type declaration statement**.    An INTEGER, REAL, DOUBLE PRECISION, COMPLEX, CHARACTER, LOGICAL, or TYPE(*type-name*) *statement*.

**Type parameter**. A parameter of an *intrinsic data type*.

**Type parameter values**. The values of the *type parameters* of a *data entity* of an *intrinsic data type*.

**Undefined**. For a *data object*, the property of not having a determinate value.

**Unspecified storage unit**. A unit of storage for holding a *pointer* or a *scalar object* of non-default *intrinsic type* that is not a pointer.

**Use association**. The association of *names* in different *scoping units* specified by a USE *statement*.

**Variable \***. A *data object* whose value can be *defined* and redefined during the execution of an *executable program*. It may be a *named* data object, an *array element*, an *array section*, a *structure component*, or a *substring*. [In Fortran 77, a variable was always *scalar* and named.]

**Vector subscript**. A *section subscript* that is an integer *expression* of *rank* one.

# Appendix E. SOLUTIONS TO EXERCISES

*Note:* A few exercises have been left to the reader.

## Chapter 2

**1.**

| | |
|---|---|
| B is less than M | true |
| 8 is less than 2 | false |
| * is greater than T | not determined |
| $ is less than / | not determined |
| blank is greater than A | false |
| blank is less than 6 | true |

**2.**

| | | |
|---|---|---|
| | X = Y | correct |
| 3 | A = B+C ! Add | correct, with commentary |
| | WORD = 'String' | correct |
| | A = 1.0; B = 2.0 | correct |
| | A = 15. ! Initialize A; B = 22. ! and B | |
| | | incorrect (embedded commentary) |
| | SONG = "Life is just& | correct, initial line |
| | & a bowl of cherries" | correct, continuation |
| | CHIDE = 'Waste not, | incorrect, trailing & missing |
| | want not!' | incorrect, leading & missing |
| 0 | C(3:4) = 'UP" | incorrect (invalid statement label; invalid form of character constant) |

**3.**

| | | | |
|---|---|---|---|
| -43 | integer | 'WORD' | character |
| 4.39 | real | 1.9-4 | not legal |
| 0.0001E+20 | real | 'STUFF & NONSENSE' | character |
| 4 9 | not legal | (0.,1.) | complex |
| (1.E3,2) | complex | 'I CAN''T' | character |
| '(4.3E9, 6.2)' | character | .TRUE._1 | legal logical provided KIND=1 available |
| E5 | not legal | 'SHOULDN' 'T' | not legal |
| 1_2 | legal integer provided KIND=2 available | "O.K." | character |
| Z10 | not legal | Z'10' | hexadecimal |

**4.**

| | | | |
|---|---|---|---|
| NAME | legal | NAME32 | legal |
| QUOTIENT | legal | 123 | not legal |
| A182C3 | legal | NO-GO | not legal |
| STOP! | not legal | BURN_ | legal |
| NO_GO | legal | LONG__NAME | legal |

**5.**

```
REAL, DIMENSION(11)     :: A    A(1), A(10), A(11), A(11)
REAL, DIMENSION(0:11)   :: B    B(0), B(9), B(10), B(11)
REAL, DIMENSION(-11:0)  :: C    C(-11), C(-2), C(-1), C(0)
REAL, DIMENSION(10,10)  :: D    D(1,1), D(10,1), D(1,2), D(10,10)
REAL, DIMENSION(5,9)    :: E    E(1,1), E(5,2), E(1,3), E(5,9)
REAL, DIMENSION(5,0:1,4) :: F   F(1,0,1), F(5,1,1), F(1,0,2), F(5,1,4)
```

Array constructor: (/ (I, I = 1,11) /)

**6.**

```
C(2,3)      legal       C(4:3)(2,1)   not legal
C(6,2)      not legal   C(5,3)(9:9)   legal
C(0,3)      legal       C(2,1)(4:8)   legal
C(4,3)(:)   legal       C(3,2)(0:9)   not legal
C(5)(2:3)   not legal   C(5:6)        not legal
C(5,3)(9)   not legal   C(,)          not legal
```

**7.**

a)
```
TYPE VEHICLE_REGISTRATION
    CHARACTER(LEN=3)    LETTERS
    INTEGER             DIGITS
END TYPE VEHICLE_REGISTRATION
```

b)
```
TYPE CIRCLE
    REAL                        RADIUS
    REAL, DIMENSION(2) ::   CENTRE
END TYPE CIRCLE
```

c)
```
TYPE BOOK
    CHARACTER(LEN=20)                   TITLE
    CHARACTER(LEN=20), DIMENSION(2) :: AUTHOR
    INTEGER                             NO_OF_PAGES
END TYPE BOOK
```

Derived type constants:

```
VEHICLE_REGISTRATION('PQR', 123)
CIRCLE(15.1, (/ 0., 0. /))
BOOK("Pilgrim's Progress", (/ 'John', 'Bunyan' /), 250 )
```

**8.**

```
T               array       T(4)%VERTEX(1)  scalar
T(10)           scalar      T(5:6)          array
T(1)%VERTEX     array       T(5:5)          array (size 1)
```

**9.**

a)
```
INTEGER, PARAMETER ::   TWENTY = SELECT_INT_KIND(20)
INTEGER (KIND = TWENTY) COUNTER
```

b)
```
INTEGER, PARAMETER :: HIGH = SELECT_REAL_KIND(12,100)
REAL(KIND = HIGH) BIG
```

c)
```
CHARACTER(KIND=2) SIGN
```

## Chapter 3

**1.**

```
A+B             valid      -C                        valid
A+-C            invalid    D+(-F)                    valid
(A+C)**(P+Q)    valid      (A+C)(P+Q)                invalid
-(X+Y)**I       valid      4.((A-D)-(A+4.*X)+1)      invalid
```

**2.**

```
C+(4.*F)
((4.*G)-A)+(D/2.)
A**(E**(C**D))
((A*E)-((C**D)/A))+E
(I .AND. J) .OR. K
((.NOT. L) .OR  ((.NOT. I) .AND. M)) .NEQV. N
((B(3).AND.B(1)).OR.B(6)).OR.(.NOT.B(2))
```

**3.**

```
3+4/2   = 5      6/4/2    = 0
3.*4**2 = 48.    3.**3/2  = 13.5
-1.**2  = -1.    (-1.)**3 = -1.
```

**4.**

```
ABCDEFGH
ABCD0123
ABCDEFGu            u - unchanged
ABCDbbuu            b - blank
```

**5.**

```
.NOT.B(1).AND.B(2) valid   .OR.B(1)                 invalid
B(1).OR..NOT.B(4)  valid   B(2)(.AND.B(3).OR.B(4))  invalid
```

**6.**

```
D .LE. C        valid      P .LT. T > 0.            invalid
X-1 /- Y        valid      X+Y < 3 .OR. > 4.        invalid
D.LT.C.AND.3.0  invalid    Q.EQ.R .AND. S>T         valid
```

**7.**

```
    a) 4*L

    b) B*H/2.

    c) 4./3.*PI*R**3     (assuming PI has value π)
```

**8.**

```
INTEGER N, ONE, FIVE, TEN, TWENTY_FIVE
TWENTY_FIVE - (100-N)/25
TEN         = (100-N-25*TWENTY_FIVE)/10
FIVE        = (100-N-25*TWENTY_FIVE-10*TEN)/5
ONE         = 100-N-25*TWENTY_FIVE-10*TEN-5*FIVE
```

**9.**

```
A = B + C      valid
C = B + 1.0    valid
D = B + 1      invalid
R = B + C      valid
A = R + 2      valid
```

**10.**

```
A = B        valid     C = A(:,2) + B(5,:5)   valid
A = C+1.0    invalid   C = A(2,:) + B(:,5)    invalid
A(:,3) = C   valid     B(2:,3) = C + B(:5,3)  invalid
```

*Chapter 4*

**1.**

```
INTEGER I, J, K, TEMP
INTEGER, DIMENSION(100) :: REVERSE
DO I = 1,100
    REVERSE(I) = I
END DO
READ *, I, J
DO K= I, I+(J-I-1)/2
    TEMP = REVERSE(K)
    REVERSE(K) = REVERSE(J-K+I)
    REVERSE(J-K+I) = TEMP
END DO
END
```

*Note:* A simpler method for performing this operation will become apparent in Section 6.10.

**2.**

```
INTEGER LIMIT, F1, F2, F3
READ *, LIMIT
F1 = 1
IF (LIMIT.GE.1) PRINT *, F1
F2 = 1
IF (LIMIT.GE.2) PRINT *, F2
DO I = 3, LIMIT
    F3 = F1+F2
    PRINT *, F3
    F1 = F2
    F2 = F3
END DO
END
```

**6.**

```
REAL X
DO
    READ *, X
    IF (X.EQ.-1.) THEN
        PRINT *, 'Input value -1. invalid'
    ELSE
        PRINT *, X/(1.+X)
        STOP
    END IF
END DO
END
```

**7.**

```
TYPE(ENTRY), POINTER :: FIRST, CURRENT, PREVIOUS
CURRENT => FIRST
IF (CURRENT%INDEX == 10) THEN
    FIRST => FIRST%NEXT
ELSE
    DO
        PREVIOUS => CURRENT
        CURRENT => CURRENT%NEXT
        IF (CURRENT%INDEX == 10) EXIT
    END DO
    PREVIOUS%NEXT => CURRENT%NEXT
END IF
```

## Chapter 5

**1.**

```
SUBROUTINE CALCULATE(X, N, MEAN, VARIANCE, OK)
    INTEGER I, N
    REAL, DIMENSION(N) :: X
    REAL MEAN, VARIANCE
    LOGICAL OK
    MEAN = 0.
    VARIANCE = 0.
    OK = N > 1
    IF (OK) THEN
        DO I = 1, N
            MEAN = MEAN + X(I)
        END DO
        MEAN = MEAN/N
        DO I = 1, N
            VARIANCE = VARIANCE + (X(I) - MEAN)**2
        END DO
        VARIANCE = VARIANCE/(N-1)
    END IF
END
```

*Note:* A simpler method will become apparent in Chapter 8.

**2.**

```
    SUBROUTINE MATRIX_MULT(A, B, C, I, J, K)
       INTEGER I, J, K, L, M, N
       REAL, DIMENSION(I,J) :: A
       REAL, DIMENSION(J,K) :: B
       REAL, DIMENSION(I,K) :: C
       C(1:I, 1:K) = 0.
       DO N = 1, K
          DO L = 1, J
             DO M = 1, I
                C(M, N) = C(M, N) + A(M,L)*B(L, N)
             END DO
          END DO
       END DO
    END
```

**3.**

```
    SUBROUTINE SHUFFLE(CARDS)
       INTEGER, DIMENSION(52) :: CARDS
       INTEGER LEFT, CHOICE, I, TEMP
       REAL R
       CARDS = (/ (I, I=1,52) /)        ! Initialize deck
       DO LEFT = 52,1,-1                ! Loop over number of cards left
          CALL RANDOM_NUMBER(R)         ! Draw a card
          CHOICE = R*LEFT + 1           !    from remaining possibilities
          TEMP = CARDS(LEFT)            !    and swap with last
          CARD(LEFT) = CARDS(CHOICE)    !    one left
          CARDS(CHOICE) = TEMP
       END DO
    END
```

**4.**

```
    CHARACTER FUNCTION EARLIEST(STRING, LEN)
       CHARACTER (LEN) STRING
       INTEGER J
       IF (LEN.LE.0) THEN
          EARLIEST = ''
       ELSE
          EARLIEST = STRING(1:1)
          DO J = 2, LEN
             IF (STRING(J:J).LT.EARLIEST) EARLIEST = STRING(J:J)
          END DO
       END IF
    END
```

**5.**

```
SUBROUTINE SAMPLE
   REAL R, L, V, PI
   PI = ACOS(-1.)
   :
   R = 3.
   L = 4.
   V = VOLUME(R, L)
   :
CONTAINS
   FUNCTION VOLUME(RADIUS, LENGTH)
      REAL RADIUS, LENGTH, VOLUME
      VOLUME = PI*RADIUS**2*LENGTH
   END FUNCTION VOLUME
END SUBROUTINE SAMPLE
```

**7.**

```
MODULE STRING_TYPE
   TYPE STRING
      INTEGER LENGTH
      CHARACTER(LEN=80)   :: STRING_DATA
   END TYPE STRING
   INTERFACE ASSIGNMENT(=)
      MODULE PROCEDURE C_TO_S_ASSIGN, S_TO_C_ASSIGN
   END INTERFACE
   INTERFACE LEN
      MODULE PROCEDURE STRING_LEN
   END INTERFACE
   INTERFACE OPERATOR(//)
      MODULE PROCEDURE STRING_CONCAT
   END INTERFACE
CONTAINS
   SUBROUTINE C_TO_S_ASSIGN(S, C)
      TYPE (STRING)      :: S
      CHARACTER(LEN=*)   :: C
      S%STRING_DATA = C
      S%LENGTH = LEN(C)
   END SUBROUTINE C_TO_S_ASSIGN
   SUBROUTINE S_TO_C_ASSIGN(C, S)
      TYPE (STRING)      :: S
      CHARACTER(LEN=*)   :: C
      C = S%STRING_DATA(1:S%LENGTH)
   END SUBROUTINE S_TO_C_ASSIGN
   FUNCTION STRING_LEN(S)
      INTEGER STRING_LEN
      TYPE(STRING) :: S
      STRING_LEN = S%LENGTH
   END FUNCTION STRING_LEN
   FUNCTION STRING_CONCAT(S1, S2)
      TYPE (STRING)      :: S1, S2, STRING_CONCAT
      STRING_CONCAT%STRING_DATA = S1%STRING_DATA // &
         S2%STRING_DATA
      STRING_CONCAT%LENGTH = S1%LENGTH + S2%LENGTH
   END FUNCTION STRING_CONCAT
END MODULE STRING_TYPE
```

*Note:* The intrinsic LEN function, used in SUBROUTINE C_TO_S_ASSIGN, is first described in Section 8.6.

## *Chapter 6*

**1.**

```
 i)      A(1, :)
ii)      A(:, 20)
iii)      A(2:50:2, 2:20:2)
iv)      A(50:2:-2, 20:2:-2)
 v)      A(1:0, 1)
```

**2.**

```
     WHERE (Z.GT.0) Z = 2*Z
```

**3.**

```
     INTEGER, DIMENSION(16) :: J
```

**4.**

```
     W       explicit-shaped
     A, B    assumed-shape
     D       pointer
```

**5.**

```
     REAL, POINTER :: X(:, :, :)
     X => TAR(2:10:2, 2:20:2, 2:30:2)%DU(3)
```

**6.**

```
     LL = LL + LL
     LL = MM + NN + N(J:K+1, J:K+1)
```

**7.**

```
     INTEGER I, J
     INTEGER, DIMENSION(100) :: REVERSE
     REVERSE = (/ (I, I=1, 100) /)
     READ *, I, J
     REVERSE(I:J) = REVERSE(J:I:-1)
     END
```

## *Chapter 7*

**1.**

```
 i) INTEGER, DIMENSION(100) :: BIN
ii) REAL(SELECTED_REAL_KIND(6, 4)), DIMENSION(0:20, 0:20) :: IRON_TEMPERATURE
iii) LOGICAL, DIMENSION(20) :: SWITCHES
iv) CHARACTER(LEN=70), DIMENSION(44) :: PAGE
```

**2.**

value of I is 3.1, but may be changed
value of I is 3.1, but may not be changed

**3.**

```
i) INTEGER, DIMENSION(100) :: I=(/ (0, K=1, 100) /)
ii) INTEGER, DIMENSION(100) :: I=(/ (0, 1, K=1, 50) /)
iii) REAL, DIMENSION(10, 10) :: X=RESHAPE( (/ (1.0, K=1, 100) /), (/10, 10/) )
iv) CHARACTER(LEN=10) :: STRING = '0123456789'
```

*Note:* the RESHAPE function will be met in Section 8.13.3.

**4.**

|       | MOD           | OUTER | INNER   | FUN     |
|-------|---------------|-------|---------|---------|
| A-B   | CHARACTER(10,2) | -     | -       | -       |
| C,D,E | REAL          | -     | -       | -       |
| F     | REAL          | -     | -       | REAL    |
| G,H   | REAL          | -     | -       | -       |
| I-N   | INTEGER       | -     | -       | -       |
| O-W   | REAL          | -     | -       | -       |
| X     | REAL          | -     | -       | REAL    |
| Y     | REAL          | -     | -       | -       |
| Z     | REAL          | -     | COMPLEX | COMPLEX |

**5.**

```
i)    TYPE(PERSON) BOSS = PERSON('SMITH', 48.7, 22)
```

ii) (a) This is impossible because a pointer component cannot be a constant.

```
    (b)   TYPE(ENTRY) CURRENT
          DATA CUURENT%VALUE, CURRENT%INDEX /1.0, 1/
```

**6.** The following are not:

iv). because of the real exponent, and
viii). because of the pointer component.

## Chapter 8

**1.**

```
   PROGRAM QROOTS          ! Solution of quadratic equation
!
      REAL A, B, C, D, X1, X2
!
      READ(*, *) A, B, C
      WRITE( *, *) ' A = ', A, 'B = ', B, 'C = ', C
      IF (A == 0.) THEN
         IF (B /= 0.) THEN
            WRITE(*, *) ' Linear: X = ', -C/B
         ELSE
            WRITE(*, *) ' No roots!'
         ENDIF
      ELSE
         D = B**2 - 4.*A*C
         IF (D < 0.) THEN
            WRITE (*, *) ' Complex', -B/(2.*A), '+-',SQRT(-D)/(2.*A)
         ELSE
            X1 = -(B + SIGN(SQRT(D), B))/(2.*A)
            X2 = C/(X1*A)
            WRITE(*, *) ' Real roots', X1, X2
         ENDIF
      ENDIF
   END
```

*Historical note:* A similar problem was set in one of the first books on Fortran programming — *A FORTRAN Primer* by E. Organick (Addison-Wesley, 1963). It is interesting to compare Organick's solution, written in FORTRAN II, on p. 122 of that book, with the one above. (It is reproduced in the *Encyclopedia of Physical Science & Technology* (Academic Press, 1987), vol. 5, p. 538.)

**2.**

```
      SUBROUTINE CALCULATE(X, MEAN, VARIANCE, OK)
         REAL X(:), MEAN, VARIANCE
         LOGICAL OK
         OK = SIZE(X) > 1
         MEAN = SUM(X)/SIZE(X)
         VARIANCE = SUM((X-MEAN)**2)/(SIZE(X)-1)
      END
```

## Chapter 9

**1.**

```
      a) PRINT '(T1, 10F6.1)', GRID
      b) PRINT '(" ", 25I5)', (LIST(I), I = 1, 49, 2)
      or                      LIST(1:49:2)
      c) PRINT '(" ", 2A12)', TITLES
      d) PRINT '(T1, 5EN15.10)', POWER
      e) PRINT '(T1, 10L2)', FLAGS
      f) PRINT '(5(" (", 2F6.1, ")"))', PLANE
```

**2.**

```
CHARACTER, DIMENSION(3,3) :: TIC_TAC_TOE
:
WRITE(NUNIT, '(T1, 3A2)' ) TIC_TAC_TOE
```

**4.**

```
(a) READ(*, *) GRID
1.0 2.0 3.0 4.0 5.0 6.0 7.0 8.0 9.0 10.0

(b) READ(*, *) LIST(1:49:2)
25*1

(c) READ(*, *) TITLES
DATA TRANSFER

(d) READ(*, *) POWER
1.0 1.EN-03

(e) READ(*, *) FLAGS
T F T F T F F T F T

(f) READ(*, *) PLANE
(0.0, 1.0),(2.3, 4)
```

**5.**

```
   CHARACTER FUNCTION GET_CHAR(UNIT)
      INTEGER UNIT
10    READ(UNIT, '(A1)', ADVANCE='NO', EOR=10) GET_CHAR
      RETURN
   END
```

# Index

## A